MW00450392

The COW

A Natural & Cultural History

First published in the United States and Canada in 2021 by
Princeton University Press
41 William Street
Princeton, New Jersey 08540
press.princeton.edu

Library of Congress Control Number: 2020945302

ISBN: 978-0-691-19870-5

Conceived, designed, and produced by
Ivy Press
an imprint of The Quarto Group
The Old Brewery, 6 Blundell Street
London N7 9BH, United Kingdom
T (0) 20 7700 6700
www.quartoknows.com

Publisher James Evans
Editorial Director Tom Kitch
Art Director James Lawrence
Commissioning Editor Kate Shanahan
Project Editor Katie Crous
Design JC Lanaway
Picture Research Sharon Dortenzio
Illustrator John Woodcock

Printed in Singapore

10 9 8 7 6 5 4 3 2 1

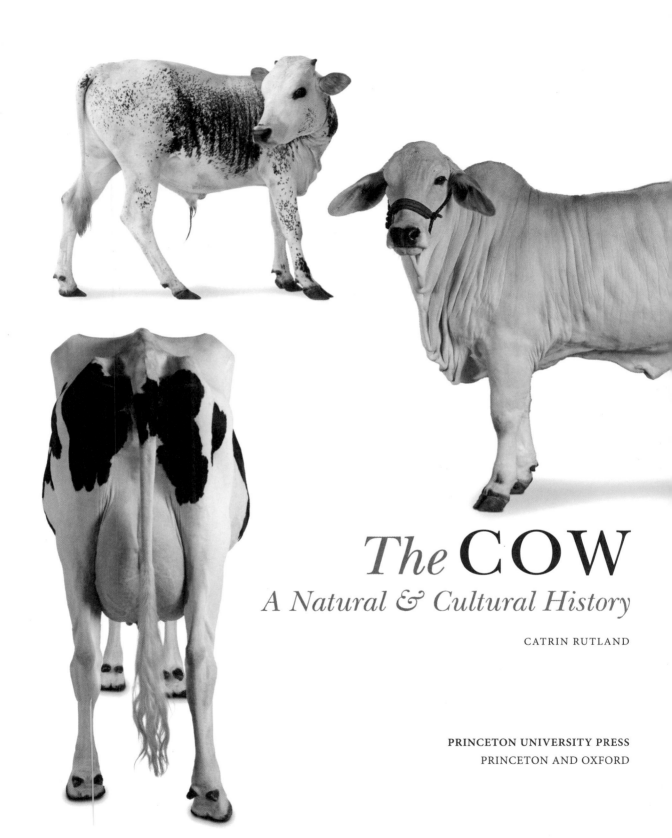

The COW
A Natural & Cultural History

CATRIN RUTLAND

PRINCETON UNIVERSITY PRESS

PRINCETON AND OXFORD

Contents

Introducing the Cow ✦

Today, ancient, historic, and very new breeds of cattle are found throughout the world, and they have all played a major part in society over the years. As we shall see in the pages to come, modern cattle breeds have descended from the now extinct auroch and the first domesticated cattle. Aurochs interacted with people, and slowly cattle became domesticated, which is shown in cave paintings through to the first farming techniques. Some modern breeds now dominate the planet, while others have become extinct or are presently endangered. Understanding their close relatives from an evolutionary point of view helps us to understand how and why cattle have adapted or have died out.

Although cattle are common throughout the world, some breeds are so rare they are officially classified as endangered. This may be because they do not produce large quantities of milk or beef, or because their breed numbers have been limited in number for other reasons. Many groups across the world work to preserve the species, and some even look at crossbreeding the original auroch genes back into the cattle genome, to create a modern breed that more closely resembles the ancient creatures. The genome sequence of modern cattle was published in 2009 and included information about 22,000 genes. Knowledge increases daily about both modern and extinct cattle, which can inform us about breeding, developing new types of cows, and quite possibly bringing back lost characteristics in some cattle. Over the years, farmers have taken great care with breeding in a bid to increase milk yields, improve beef quality, or to aid survival in difficult climates. As a result, some breeds have become more resistant to disease, while others can now cope with high temperatures and periods of drought. Combining traditional breeding knowledge with advancing techniques such as genetics and artificial insemination is helping to reduce disease incidence and target specific traits or disorders in herds.

Left: *As the ancient auroch became domesticated, our ancestors drew pictures of their animals. This prehistoric art in the Lascaux IV caves, France, gives us a glimpse into this past.*

THE IMPORTANCE OF CATTLE TO HUMANS

From the first domestication events to the present day, cattle have become increasingly important to people. They provide us with milk, meat, leather, and even their waste products are used to build houses and fuel fires. India now has the highest number of cattle in the world, and also has the highest number of people following vegan and vegetarian diets. Therefore, to many, beef is not important, and their religion tells us why cattle are so revered. Hinduism is widely followed in India, and cattle are treated as sacred animals rather than used for their meat.

Some countries have very few cattle, usually because the climate is inconducive to keeping large numbers of these large herbivores. Greenland presently has very few cattle, but when the Vikings arrived in the tenth century, the island was covered in vegetation, so these seafaring Scandinavians successfully settled on the land, farmed, and kept cattle. As the climate and vegetation changed, cattle became more difficult to rear, so today the numbers are reduced, but the population still values its cattle.

With more than a billion cattle on the planet, the importance of cows extends well beyond food production. These animals have been a key species in moving people away from the hunter-gatherer system into the agriculture-dependent way of living for so many societies throughout the world. Cattle have worked the fields alongside farmers, pulled carts, and provided the fields with fertilizer. They have provided leather for clothing, shoes, and other essential items, and have even been used as decorative ornaments, and as sporting and show animals. Bovines have also played important roles in culture and religion—children grow up learning, singing, and reading about these important animals in their everyday lives.

Right: *Zebu cattle originated in South Asia from Indian aurochs and have the characteristic fatty hump and dropped ears. Providing meat and dairy products, in addition to dung for fuel and manure and hides for clothes, shoes, ornaments, and household goods, their interactions with people have a rich history.*

THE ECONOMY OF THE COW

The major role of cow-related products in food production inevitably attaches an economic value to cattle. People in most countries will associate the cow with milk production. For years cows have produced milk for their calves, but increasingly the cow is bred to provide milk for humans. Whether that is milk for our breakfasts, in our recipes, or as feed for infants, it is a vital economy. Understanding the anatomy and physiology of milk production is essential to fully appreciate the economic value of cattle.

Over the years the anatomy of some breeds has changed greatly in order to produce vast quantities of milk. Milk is produced via the mammary glands, which are actually modified sweat glands. Cows have also developed the ligaments and strong tissue needed to contain the heavy weights of milk in the udder, and for the whole body to cope with the demands made in order to keep up with milk production. Particular breeds have adapted and been bred for high milk yields, and the economic ramifications of this have huge impacts across the globe. In addition, a number of illnesses and diseases are directly related to milk production, therefore much care has to be taken to understand and treat these conditions.

Beef production is the other major economic drive of dairy cattle. Meat production has consistently risen in the last few years, with market and consumer changes in demand for the types and quality of meat desired. Balancing feed efficiency, calf size, ease of calving, growth rates, and leanness, along with maximizing milk production, is no easy task.

Top: *With 7 billion dairy consumers and around 210 gallons of milk produced from cows annually, milk is an international commodity.*

Above: *Nearly 28 billion pounds of beef are produced annually in the USA alone.*

ANATOMY & BEHAVIOR

The male and female reproductive systems are essential in production animals, and high fertility rates must be maintained. The average pregnancy in cows is nine months, just like people; however, the cow's needs throughout pregnancy and birth differ from that of a human female. Cattle also have very unique skeletons, organs, cardiovascular systems, genetics, and senses when compared to other animals. In order to care fully for bovines, diseases and disorders must be understood, herd health and welfare must be maintained, and food, water, facilities, and behavior must be taken into consideration—especially with around a billion cattle on the planet, each chewing 50 times a minute, eating around 100 lb (45 kg) of feed each a day.

Cattle behavior is a very interesting topic, given the notorious attitudes of some bulls, and indeed cows, and their memory and intelligence. Studies have shown that milk yield can increase when cows have music played to them, but only if the music is slow and rhythmic, as no increase was seen when fast music was played. It also appears that reading Shakespeare aloud to cows can increase milk yield! Despite thousands of years of observing and studying cattle, there is still much to learn about their behavior and needs.

More than any other time in history, the discussion around the environment and climate is key to the future role of cattle. Farming and breeding practices in relation to preserving land—and our planet—are being questioned, challenged, and discussed. However, from religion and culture through to petting farms and art, our love for cattle crosses the globe, and this book offers further insight into our relationship—past, present, and future—with these varied and revered animals.

Right: *Petting farms enable children and adults alike to have close contact with cows and their young, encouraging a positive relationship between bovine and human.*

CHAPTER 1

Evolution & Development

Ancestors of the Modern Cow ❧

Modern cattle have a remarkable history. Starting with their classification as mammals, they are within the order *Artiodactyla*, then the infraorder *Pecora*, and finally the family *Bovidae*. Within the family there is a very wide variety of species so the family is broken down further into eight subfamilies. Modern cattle and their closest relatives are in the *Bovinae* subfamily, which splits further into three tribes.

Modern cattle evolved from the aurochs, a now-extinct species of wild cattle found throughout Europe, North Africa, and much of Asia. They are also known as aurochsen, aurochses, urus, or ure, and the Latin classification is *Bos primigenius* (but has previously been *Bos urus* and even *Bos taurus*). There were three subspecies of *Bos primigenius*, which are now all extinct, and two domesticated subspecies. The extinct aurochs were the Eurasian aurochs (*B. p. primigenius*), the Indian aurochs (*B. p. namadicus*), and the North African aurochs (*B. p. africanus*). The aurochs belong in the same tribe as cattle, the *Bovini* tribe.

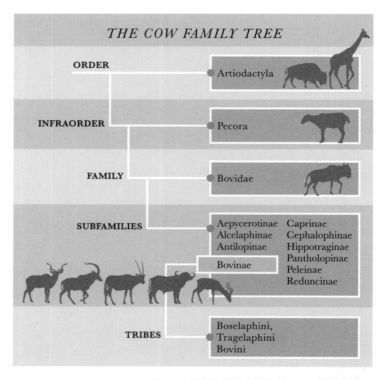

THE COW FAMILY TREE

ORDER — Artiodactyla

INFRAORDER — Pecora

FAMILY — Bovidae

SUBFAMILIES — Aepycerotinae Caprinae
Alcelaphinae Cephalophinae
Antilopinae Hippotraginae
Bovinae Pantholopinae
Peleinae
Reduncinae

TRIBES — Boselaphini,
Tragelaphini
Bovini

Right: *The wild cattle of Chillingham are an ancient breed dating back over 800 years. Descending from ancient wild cattle in Britain, their earliest ancestors are the auroch.*

Left: Wild aurochs roamed the planet 2 million years ago, with the last of the three subspecies becoming extinct in 1627.

According to archeological remains, the Indian aurochs were the first to evolve around 2 million years ago, and the last known individual died about 4,400 years ago in south India. Prior to extinction they migrated both east and west, reaching Europe 270,000 years ago. The Indian subspecies was also domesticated and gave rise to the indicine (zebu) cattle around 9,000 years ago. It is interesting to note that these were living near the desert and, therefore, have likely influenced the modern zebu, which is hardy during times of drought.

The Eurasian aurochs survived for the longest of the three subspecies, living across Europe, Siberia, and Central and East Asia. Domestication of the Eurasian subspecies around 6000 BCE in the Middle East and potentially the Far East, too, gave rise to modern taurine cattle. Many of the features of these ancient breeds can be seen in modern cattle: the horn shape and the dark coloration in bulls with the lighter eel stripe along the back, for example. The aurochs were popular in Roman arenas. These ancient herbivores were also hunted; as they became rare they were hunted only by the nobility and royal families, when the illegal killing of an auroch could result in the death sentence. As well as being hunted, a lack of grazing land and the contraction of disease from other domesticated cattle led to the reduction of aurochs, until they were found only in Europe. They went extinct in the seventeenth century, with the last known female dying a natural death around 1627 in a forest in Mazovia, Poland.

Below: Zebu cattle, also known as indicine or humped cattle, descended from Indian aurochs. Today more than 75 zebu breeds exist, tolerating heat, drought, and sunlight well.

The North African auroch descended from the Eurasian species coming in from the Middle East around 25,000 years ago. It is thought that it died out before the Middle Ages and was probably not domesticated, unlike the other two subspecies. There is some debate over this last point, though, as Sanga cattle may have originated from these aurochs, which later bred with taurines and zebus. The Sanga has a hump like the zebu, but genetic studies show significant similarities with the taurines. There has also been much debate as to whether the Turano-Mongolian cattle presently living in China, Mongolia, Korea, and Japan may have been domesticated from the aurochs around 35,000 years ago, but investigations are still underway to determine their genetic background. It is also thought that the European bison originated from the aurochs.

These ancient aurochs were large, ranging from 1,540 lb (700 kg) to 3,310 lb (1,500 kg), similar to the European bison, and with horns reaching up to 32 in. (80 cm) long. As with modern cattle, there was a large difference in size between the males and females. For example, male aurochs from Denmark and Germany reached shoulder heights of 61–71 in. (155–180 cm); females were generally 10 in. (25 cm) smaller. Examination of their bones shows females were around 10–20 percent smaller, but this depended on location, food availability, the individuals themselves, and their age, in addition to other factors. In modern times, people have tried to create equivalents of the aurochs by breeding traditional cattle (see The Cow Genome, page 78). The breed known as the Heck was created by zoo directors Heinz and Lutz Heck in Germany in the 1920s. Even in the last decade, a number of programs have been established to research, understand, and recreate the auroch.

Below: *The auroch skull is larger and more elongated than most modern cattle breeds. Its impressive horns were used to kill predators such as wolves and bears.*

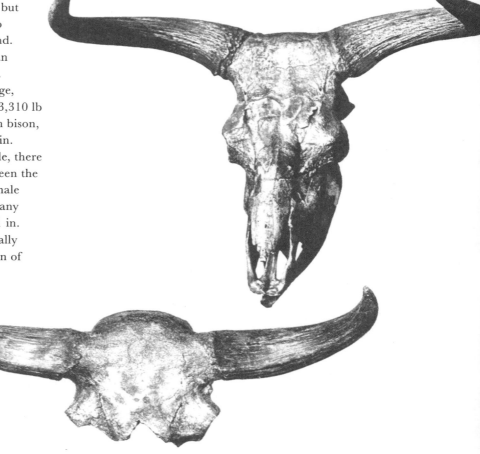

If we turn the clock back even farther, it is thought that *Bos acutifrons* was the ancestor of the aurochs. Fossil excavations of remains, including horns, from Pakistan and India dated them to the middle of the Pleistocene period, 2.5 million to 11,700 years ago. The aurochs came into existence around 2 million years ago, and lived for around 500,000 years with their ancestors, but the *Bos acutifrons* became extinct around 1.5 million years ago. Ancient members of *Equus* (a genus which now includes horses, asses, and zebras) and *Elephas* (including elephants) would have roamed these lands at around the same time.

Although recent research points to a lineage from modern cattle to the aurochs, and before that the *Bos acutifrons*, older studies indicated that the ascendancy may have included features from gaur, banteng, bison, yak, and kouprey. Prior to *Bos acutifrons*, another tribe from the *Bovinae* subfamily gave rise to the *Bovini* tribe, to which all the previous species belong. This tribe is called *Boselaphini*, which includes antelopes and still has two genera alive today: *Nigali* (blue bull) and *Tetracerus quadricornis* (four-horned antelope). The oldest fossils and now extinct members date back as far as 18 million years ago. Research suggests that *Boselaphini* were the ancestors of both *Bovini* (including modern cattle) and *Tragelaphini* (the other tribe in the *Bovinae* subfamily).

Below: *This Spanish cave painting is believed to show an auroch from the Paleolithic period. The ancestor to modern cattle, it was found across Europe and several other countries, and was depicted in many types of art.*

Related Species & Distant Cousins ✑

By looking at the similarities, differences, general characteristics, and attributes of the three tribes in the *Bovinae* subfamily, we can get a better understanding of cattle, their closest relatives, and their distant cousins.

BOSELAPHINI

The *Boselaphini* are ancestors of *Bovini* and *Tragelaphini*. The tribe includes many extinct members but the living species are the *Tetracerus* and the *Boselaphus*. The modern antelope *Tetracerus* stands at around $1^1/2$ ft (0.5 m) tall, lives in India and Nepal, and differs from most bovids in having four horns instead of two. It is a grazer that lives mostly in solitude or in very small groups. The low numbers left alive mean it is a threatened animal, and it is classified as vulnerable. Not only is its natural habitat being threatened, but also, because of its distinctive horns, it is valued by hunters.

The last remaining *Boselaphus*, commonly called the nilgai or blue bull, is *Boselaphus tragocamelus*. Its name is a combination of Latin and Greek words meaning cow/ox, deer, he-goat, and camel, which gives an insight into the many different physical attributes of this bovid. In contrast to the small *Tetracerus*, this antelope stands at 3–5 ft (1–1.5 m) in height. It also lives in India and Nepal, in addition to Pakistan, and recently was seen in Bangladesh. In the 1920s and 30s, these animals were introduced into Texas, and present day numbers show around 37,000 of them running feral in that state, with more found in other states, such as Alabama, Florida, and Mississippi. Individuals generally have a tame nature, with calves being shy, and the population is larger than that of the *Tetracerus*. Sadly, these animals can cause damage to crops, which has led to some states classifying them as vermin; however, they are not endangered at the moment.

Below: *The last remaining Boselaphus is the nilgai. This shy bovid is the largest antelope in Asia. Archeological remains indicate that people were hunting them 8,000 years ago.*

BOVINI

Usually split into three subtribes, *Bovini* include: 1) soala; 2) African buffalo, anoa, and wild water buffalo; and 3) bison, taurine, and Asiatic cattle (including domesticated cattle). Although cattle are often referred to as being domesticated, it is important to acknowledge that other species of wild bovines have been domesticated too, such as the gaur, wild yak, banteng, and the wild water buffalo. So, despite the word "wild" appearing in their names, some have been used by humans.

The *Bovini* tribe is broken down into several genus groups. The first such group is *Bubalus*, which includes species such as water, swamp, and river buffalo; lowland and mountain anoa; tamaraw and the extinct Cebu tamaraw. It is fair to say that the anoa and tamaraw both look like their commonly used names of the mini and dwarf buffalo, while the Cebu tamaraw would have been even smaller in height, at around $2^1/2$ ft (75 cm), and weighing just 350 lb (160 kg). The buffalo, meanwhile, is one of the largest bovines and can reach $4^1/2$ ft (133 cm) tall and weigh over 2,200 lb (1,000 kg). Domestication started around 5,000 years ago in India for the river buffalo, and 4,000 years ago for the swamp type in China. Although many are domesticated, there are still feral herds present in many countries. Modern buffalo probably descended from the wild water buffalo.

Despite the large numbers of buffalo that exist today, this wild species is endangered and has been listed as such since 1996, with only around 3,400 animals left in 2010. They are usually larger than other buffalo, and this majestic creature carries the largest horns of any bovid alive today—spanning up to $6^1/2$ ft (2 m). These days, most of this endangered species live in protected parks, but many have interbred with feral or domesticated buffalo. This is one of the causes for its endangerment, but being hunted, changing environmental conditions, disease, and loss of habitat are also severe threats. Buffalo often live up to 40 years, and the older females lead the herds.

Below: *Two types of water buffalo exist today—the swamp and river buffalo. People worldwide depend on the 130 million water buffalo for food, byproducts, and draft work.*

Older males are often solitary, with younger males forming small herds of up to 10 animals. Like cattle, they are grazing animals, but, in addition to eating grasses, herbs, and shrubs, they can often be partial to crops and sugar cane, which can cause conflict with local farmers. The differing types of buffalo can interbreed and produce fertile young; however, buffalo crossed with cattle do not produce living young. The bacteria found in their rumen differs significantly from cattle's, and their milk is higher in fatty acids and proteins.

The next genus in the *Bovini* tribe is called the *Syncerus*, which includes the African buffalo. These buffalo are very distinct from those discussed above as they have not been domesticated, due predominantly to their aggressive nature. Another interesting genus is the *Pelorovis*, which only includes the extinct giant buffalo. It was one of the largest ever ruminants known, with males reaching 4,400 lb (2,000 kg) and measuring up to 9.8 ft (3 m) long. The genus *Bison* includes the American bison, the wisent, the extinct species of *Bison palaeosinensis*, steppe wisent, ancient bison, and the long-horned bison. The common names of buffalo or American buffalo reflect many of the similarities that we see between their anatomy, general looks, and behavior in comparison to true buffalo.

The genus called *Pseudoryx* includes just the saola. These bovines are related to cattle, goats, and antelopes, and generally live in Vietnam and Laos. This rare animal is often kept in captivity and is very rare in the wild; it was only officially discovered and identified as a new species in 1992. Due to a gentle nature and infrequent sightings in the wild, they are often referred to as Asian unicorns.

Last but not least, the final genus in the *Bovini* tribe is *Bos*, which is particularly important as it includes aurochs, banteng, gaur, gayal, yak, wild yak, *Bos palaeosondaicus* (extinct), kouprey, and, of course, domestic cattle, which includes both taurine cattle and zebu. The rest of this book concentrates on *Bos* and particularly modern cattle.

Left: *Close relatives of modern cattle, today's 1.5 million Banteng are domesticated; however, wild Banteng are endangered. It was the second endangered species to be cloned successfully.*

TRAGELAPHINI

The third and final tribe in the *Bovinae* subfamily is called *Tragelaphini*. It contains just two genera: *Taurotragus* (eland) and *Tragelaphus*. *Tragelaphini* and *Boselaphus* are more closely related to each other than to *Bovini*. *Tragelaphini* actually evolved from *Boselaphini* around 15–18 million years ago. It is believed that this happened in Africa, but they then spread into Eurasia and potentially South Asia. The *Tragelaphini* tribe is commonly called the spiral-horned antelopes. There are debates as to how many different genera and species exist within the tribe (thought to be between nine and 25) but they include kudus, elands, nyalas, and bushbucks. Most spiral-horned antelopes are thriving, but the giant eland and mountain nyala are now vulnerable and endangered, respectively.

THE DEPENDABLE BUFFALO

The buffalo is immensely important worldwide and has been described as the most depended upon domesticated species, with over 130 million alive in the world. Although this number is lower than cattle, they have varied uses and are essential in many countries for meat and dairy production, skin products, bones and horns used to make jewelry and musical instruments, carrying loads and farming, fighting and racing events, and religious and cultural ceremonies and festivals. The buffalo originated in South and Southeast Asia and China but is now found throughout the world, including America, Australia, Europe, and parts of Africa. The buffalo in America should not be confused with the American bison, which is often referred to as the American buffalo, or simply "buffalo."

Left: *The nocturnal forest-dwelling Bongo antelope belongs to the* Tragelaphus *genus. Males and females have spiral horns and a chestnut-colored coat with stripes.*

The Full Breadth of the Species ⤜

Below: *Modern cattle are classified into two groups: indicines, also called zebus, which have humps; and taurines, which are humpless. The two types can breed.*

Modern day cattle are based on two types of cattle which have evolved due to their traditional geographical locations: the taurine, which is humpless and originated from Europe, Africa, and Asia; and the indicine, which are humped and came from South Asia and East Africa. Both types can now be seen throughout the world. Indeed, taurine cattle can successfully breed with indicines, thereby producing many hybrids and breeds. They can also breed with other members of the *Bos* genus, such as yaks, gaur, banteng, and even bison.

By looking at *primigenius*, *indicus*, and *taurus* in turn, we can start to see their similarities and differences, and comprehend why they are distinct subspecies.

BOS TAURUS PRIMIGENIUS

Bos taurus primigenius (the aurochs) survived up until the year 1627, but all three of the subspecies are now extinct. The striking horns reaching up to 32 in. (80 cm) needed a larger skull to support them than that seen in most modern cattle. They also had more slender legs and good musculature throughout their body, especially in the neck region, and relatively small udders—even in pregnant cows. While it is difficult to put a coat color to these ancient animals, cave drawings, Egyptian grave drawings, and pieces of writing give us some clues. They indicate that the young were chestnut-colored, with males developing a darker coat and the white eel stripe, and the females retaining their red coloration. These sources also tell us that they could be aggressive when under threat, and the bulls would fight, even to the death, during mating season. These now extinct animals give us fascinating insights into modern cattle. Many scientific studies have traced the ancestry of cattle back to the auroch, and some dispute that the auroch is extinct. With auroch DNA present in so many of today's breeds, and to varying degrees, some people suggest that the auroch has evolved rather than become extinct. It is clear that there are many differences in the skeletons, but we can also see just how closely related the auroch and modern breeds are.

Below: *Modern bulls tend to have a smaller skull and can have equally impressive horns, larger legs, and a more docile temperament than their auroch ancestors.*

BOS TAURUS INDICUS

This subspecies is also known as zebu, indicine cattle, or humped cattle. In general, they can be recognized by the fatty hump on their shoulder. In some breeds, the ears will droop and they have a large dewlap—the mass of loose skin hanging from the neck or throat, sometimes called the "briefcase folds" in zebu. Originating in Asia, *indicus* are now common throughout the world. They have a good tolerance for high temperatures and drought, which makes them popular in hotter countries. This tolerance is due partly to their ability to lower their metabolic rate when nutrition is low, around 40 percent more efficiently than most taurine breeds, meaning they can survive food and water shortages better. They need less water than most taurine breeds, and can put on weight faster. In the present day, we see many different breeds that fulfill many important roles in society: producing milk and beef, working the land, pulling carts, carrying people and goods, being kept as pets, and their hides being used.

BOS TAURUS TAURUS

Also known as taurines and European taurines, *taurus* descended from aurochs in the Near East, which roughly corresponds to Western Asia, Egypt, and Turkey. Despite originating from warmer climates, these cattle have now adapted well to cooler conditions. Indeed, taurines do not handle heat as well as zebus from a very early age; studies have shown that even taurine embryos are not as capable of defending themselves against high temperatures. Taurines also have a lower resistance to ticks than their

Below: *Indicine cattle are commonly seen pulling carts and plows, and provide vital food and milk to communities throughout the world.*

Below right: *Despite originating in the Near East, taurines do not handle high temperatures or drought well in comparison to indicines, but they have adapted well to colder climates.*

close subspecies the zebu, and a lower resistance to tropical diseases. So, although we do see taurines in West Africa and Latin America, on the whole there are fewer of them in tropical countries today. Taurines are easily differentiated from zebu, as they do not have the zebu hump. There are many differences between the two subspecies but they can still breed, and hybrid offspring and breeds are common. Humans have a large role to play in modern breeding programs for cattle, but even native and older breeds were hybrids—the Damietta in Egypt and the Damascus cow in Syria are clearly hybrids from early zebu and taurines. Indeed, both types have lived in these areas for a long time.

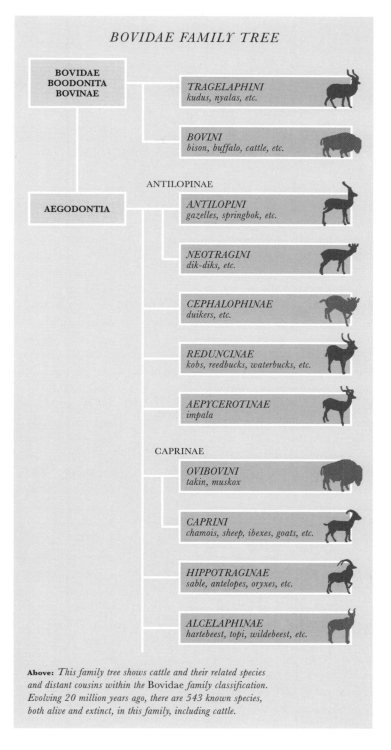

BOVIDAE FAMILY TREE

BOVIDAE BOODONITA BOVINAE

TRAGELAPHINI
kudus, nyalas, etc.

BOVINI
bison, buffalo, cattle, etc.

ANTILOPINAE

AEGODONTIA

ANTILOPINI
gazelles, springbok, etc.

NEOTRAGINI
dik-diks, etc.

CEPHALOPHINAE
duikers, etc.

REDUNCINAE
kobs, reedbucks, waterbucks, etc.

AEPYCEROTINAE
impala

CAPRINAE

OVIBOVINI
takin, muskox

CAPRINI
chamois, sheep, ibexes, goats, etc.

HIPPOTRAGINAE
sable, antelopes, oryxes, etc.

ALCELAPHINAE
hartebeest, topi, wildebeest, etc.

Above: *This family tree shows cattle and their related species and distant cousins within the* Bovidae *family classification. Evolving 20 million years ago, there are 543 known species, both alive and extinct, in this family, including cattle.*

Conquering the Globe ❧

With over a billion cattle across the world, it is fascinating to see where they came from originally, and how they evolved and then moved to different countries.

It is clear that even before domestication of the auroch, the ancestors of modern cattle, that bovines were important to humans. The aurochs were living alongside people and were being hunted; these were wild animals, not domesticated. We know of two critical time points, between 8,000 and 10,500 years ago, when the aurochs were domesticated, giving rise to the zebu and taurine cattle. We know that over time, the auroch ancestors were found throughout most of Europe (but not in Ireland or northern Scandinavia), Central Asia, East Asia (Japan), North Africa, the Middle East, and India, and remains were dated to 3,000 years ago in eastern China.

THE RISE AND MOVEMENT OF *TAURINES*

Modern DNA evidence has shed further light on where and when the taurines evolved. The auroch was domesticated in Caucasus and Mesopotamia. Situated between the Black Sea and the Caspian Sea, near Russia and Georgia, the auroch could flourish in Caucasus. It is thought that around 80 aurochs were tamed in Mesopotamia (present-day Iraq, eastern Syria, and southeastern Turkey) around 10,500 years ago, and the resulting subspecies was the taurine. During the period that Eurasian aurochs would have been domesticated in Mesopotamia, humanity was undergoing the great Neolithic Revolution in this same area. In the last stage of the Stone Age, the Neolithic period began, around 12,000 years ago, and ending around 6,500 years ago. Humans began to form settlements, and agriculture quickly developed, with the first cereal crops being planted in Mesopotamia. This would have been helped by the wheel being invented in this region and the development of writing, mathematics, and scientific studies of the skies. With all these factors combining, the domestication of the auroch into modern cattle would have helped develop these civilizations and the area to flourish.

Below: *This Neolithic carving originating from Turkey is one of the many ancient artifacts that highlight the importance of cattle to people throughout the ages.*

THE RISE AND MOVEMENT OF THE ZEBU

Around the same time as the taurines were evolving, the zebu was domesticated from Indian aurochs in South Asia. The first zebu were present 8,000–10,000 years ago, and evidence puts the zebu in Pakistan in 6000 BCE. We also know that the Indian aurochs lived alongside zebu until at least 4,000–5,000 years ago.

By 2000 BCE, zebu were in Egypt—they were drawn on rocks and tombs, and shown on pottery—but this is not where they originated, and it is thought that they were not widespread in this region. Around 1,000 years ago, the zebus were imported into East Africa, and it is thought that they came from the Horn of Africa (the region containing Djibouti, Eritrea, Ethiopia, and Somalia) or were transported by sea. Bones dated from 100–200 years BCE were found in Kenya. A bronze statue of the humped cattle was dated to around 330 CE in Ethiopia, but naturally this is more of an indication of their presence rather than proof. Archeological evidence puts the zebu as living in the Horn of Africa by the year 1000 CE. Note that this is later than other parts of Africa, but these regions have not had as many archeological digs of the type needed to identify cattle; therefore, zebu could have been present earlier in the Horn of Africa. This information combined correlates well with Egyptian writings about trade at this time. They discuss trade links between Egypt and Punt, on the coast of Ethiopia 2000–1400 BCE, and tell of "short-horned cattle" being transported in raftlike boats. After that date, we know that zebu existed in the sub-Saharan region in the years 700–1500 CE. It is thought that the opening of new trade routes facilitated these cattle being traded and imported. The first archeological bone evidence discovered in the 1940s and 1950s dates the fossils from around 1575 CE, though earlier animals are thought to have been present.

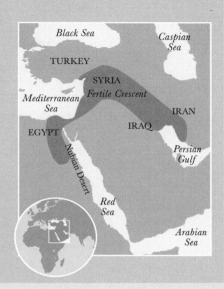

THE FERTILE CRESCENT

The Fertile Crescent, the "cradle of civilization," was home to domesticated cows, goats, sheep, and pigs.

The area was comprised of modern-day Iraq, Syria, Israel, Egypt, Palestine, Lebanon, Jordan, and parts of Turkey and Iran.

Black Sea
Caspian Sea
TURKEY
SYRIA
Fertile Crescent
Mediterranean Sea
IRAN
EGYPT
IRAQ
Nubian Desert
Persian Gulf
Red Sea
Arabian Sea

Left: *Evidence from carvings, such as this one of a zebu, cave paintings, ancient bones, historical manuscripts, and DNA have all helped us understand the rich history and evolution of cattle.*

Right: As people traveled, so did cattle. Christopher Columbus introduced cattle into the New World in 1493, and Texas Longhorns descended from those first cattle.

THE MYSTERY OF THE SANGA AND TURANO-MONGOLIAN CATTLE

It has been suggested that two other possible domestication events happened but, to date, these have not been verified as separate subspecies. The Sanga cattle came from East Africa, and we see them depicted in ancient Egyptian murals, so we know that they were transported into that region. Archeological evidence puts the cattle in South Africa in the ninth to eleventh centuries. There is much debate over whether they came from an independent domestication event, though, as it is also likely that they are a hybridization of the zebu and taurine in East Africa.

It is also possible that the Turano-Mongolian cattle arose from an independent domestication event of aurochs around 35,000 years ago. They are anatomically and genetically different to taurine cattle, and developed a tolerance to extremely cold and harsh conditions; however, they are more similar to taurines than zebu cattle. Turano-Mongolian cattle now live in Mongolia, China, Korea, and Japan.

GLOBAL DOMINATION

Once introduced to Africa, zebu were bred with taurine cattle to produce hybrid breeds. Early spread of both zebu and taurines happened across northern Africa and the Middle East. We know that the majority of the 1,000-plus breeds that exist today are taurines. In total, there are around 75 breeds of zebus originating from both Africa and India, and the remainder of the 1,000 breeds are taurines or hybrids of the two types. Cattle were transported throughout the centuries to differing parts of the world, and the number of cattle has expanded rapidly over the years. Columbus was responsible for bringing cattle to the Western Hemisphere, in 1493 reaching the Caribbean; in 1519 Hernán Cortés transported some to Mexico; while Juan Bautista de Anza introduced cattle to California in 1773. Zebu were introduced to Australia in 1788.

Records suggest that the founding two bulls and five cows were part of the "First Fleet"—the first ships to arrive from England to form the first European settlement in Australia. The ships left in 1787 and picked up the cattle in South Africa. It was not until 1832 that the Australian cattle dog was fully developed; up until this point, other dogs were used to help herd cattle, but the cattle were not very tame and were difficult to handle, resulting in the development of the Australian cattle dog. By 1814, New Zealand had cattle imported from Britain, and later from Europe.

In 2018 it was estimated that there were just over a billion cattle (including buffalo) worldwide; similar numbers were estimated in 2012. Some estimates even put the number as high as 1.4 billion animals, around 400 million of which were dairy cows. The highest numbers of cattle were present in India, Brazil, China, the United States, and then the European Union, although year on year, the numbers fluctuate. These top five regions contained 82 percent of the entire world's population, with India having about 30 percent—an astonishing 305 million cattle. Zebu were only introduced to Brazil in the twentieth century, where they were crossed with taurines, so it is fascinating that their numbers have risen so dramatically, placed second on the world population numbers of cattle and with the highest number of dairy cows, too, at over 20 million. Interestingly, the dairy populations differ from the total cattle number, although India and Brazil come in at first and second places for both categories; Sudan, followed by China then Pakistan, come in at third, fourth, and fifth places for dairy.

Far left: Man's best friend proved a good choice for cattle handling, as breeds such as Australian cattle dogs were developed. Intelligent, agile, and sturdy, they excelled at herding.

Left: Hernán Cortés (pictured) and Gregorio Villalobos introduced cattle from Cuba to Mexico. Cortés introduced branding, in the form of his three Latin crosses, to the Western Hemisphere.

Two-hundred and six countries/states/dependencies are listed as housing cattle, with Saint Pierre in the North Atlantic containing just 30 cattle, and Greenland, with a human population of around 56,000 people, potentially containing no officially listed cattle—although it is likely there are some at any one time. A few south-facing slopes on the island can provide for small numbers. With its cold winters, sparse vegetation, and an icecap covering 80 percent of the island, Greenland is not the easiest place to keep cattle, but this has not always been the case. Excavations have shown that cattle bones exist on the island, and it is likely that the Vikings kept cattle. The Vikings were good farmers and hunters, and, after arriving in the tenth century, they established over 600 farms. Greenland at that time was much warmer. By the fifteenth century, the climate was much colder as the "Little Ice Age" hit the island. Civil unrest was high, and the Vikings deserted Greenland. In a modern day twist, due to climate change, the environment is again becoming warmer and more conducive to farming. Yet sparse vegetation and extremely high costs mean that subsidies are essential to keep Greenland's farms going, and many people are moving away from the demanding and economically challenging lifestyle into other jobs.

Some other places have fewer cattle because other animals from the *Bos* genus undertake similar roles and are more acclimatized to the conditions. For example, yaks are numerous in Nepal and Tibet, the gaur and water buffalo are common in Southeast Asia, and fewer zebu and taurines reside in these regions.

This map shows where early cattle migrated to/were transported to. The size of the cattle picture depicts its relative population size in that region in comparison to the other cattle types.

Taurine: Aurochs in Mesopotamia (present-day Iraq, eastern Syria, and southeastern Turkey) were thought to have been tamed around 10,500 years ago and would later be known as taurine cattle.

Bos taurus

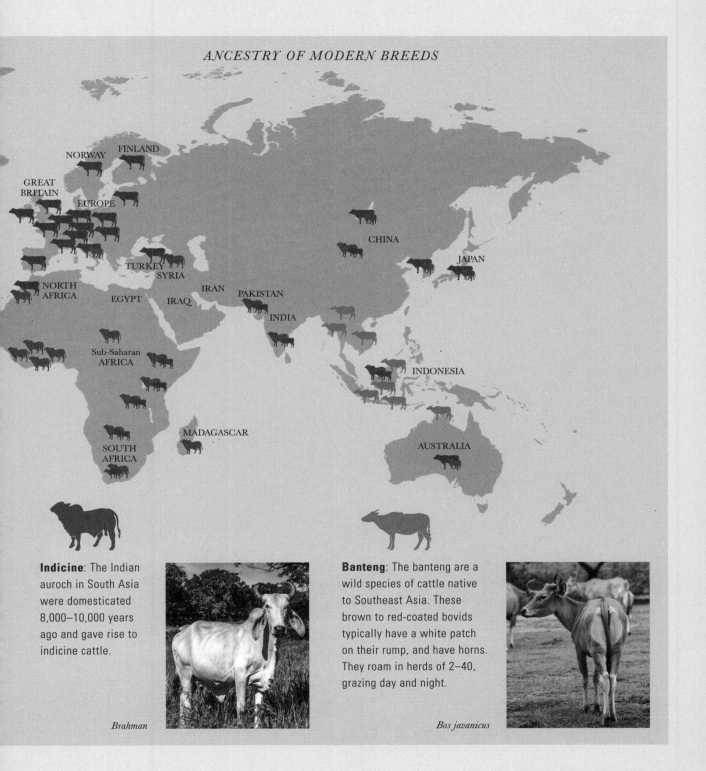

ANCESTRY OF MODERN BREEDS

NORWAY FINLAND

GREAT
BRITAIN

EUROPE

CHINA

JAPAN

TURKEY
SYRIA

NORTH
AFRICA

IRAN

PAKISTAN

EGYPT IRAQ

INDIA

Sub-Saharan
AFRICA

INDONESIA

MADAGASCAR

AUSTRALIA

SOUTH
AFRICA

Indicine: The Indian auroch in South Asia were domesticated 8,000–10,000 years ago and gave rise to indicine cattle.

Brahman

Banteng: The banteng are a wild species of cattle native to Southeast Asia. These brown to red-coated bovids typically have a white patch on their rump, and have horns. They roam in herds of 2–40, grazing day and night.

Bos javanicus

First Links with Humans ✑

The domestication events that created the taurine and zebu cattle, and potentially the Sanga and Turano-Mongolian cattle, were the first links where the relationship between humans and modern cattle became important. Cattle were being managed by humans rather than purely being hunted. The previously wild animals were now farmed, transported, and cared for by people, receiving food, water, and protection, while also being used for food themselves. Cattle were used for draft work and even for religious ceremonies. Although most of the draft work would eventually be done by horses in many countries, the importance of cattle as a food source and in religion and culture is still paramount.

Trying to piece together the geographical puzzle that exists in order to show cattle movement has itself been evolving, as evidence has been uncovered and techniques have become more advanced. Much of the evidence comes from looking at bones in archeological digs. In recent years, DNA evidence has greatly enhanced our knowledge on the types of animals, including cattle, found in particular areas. Genetic studies show the migration of both people and their cattle across the continents, and show that people took their cattle with them.

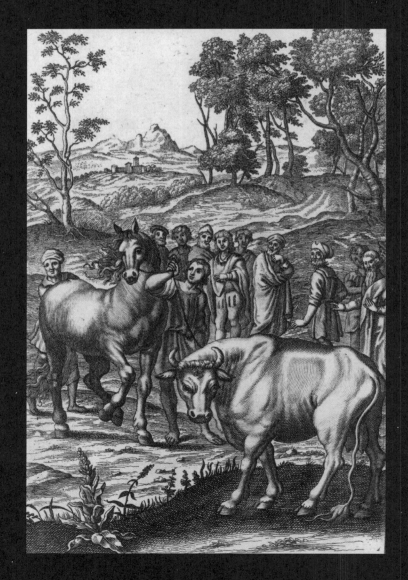

In addition, genetics can be used to ascertain relationships between breeds and understand when they developed.

Techniques such as carbon dating have gone hand in hand with archeological studies, while visual art such as sculptures, tomb and rock paintings, and other artistic forms of expression have shown us where the humped zebu and the taurines were. Literature has also helped, ranging from specific mentions about cattle through to understanding which trade routes and ship voyages were happening during different parts of history. This developed an understanding of who might be trading and transporting cattle. More unusual evidence comes from the other types of goods being traded from different regions in the world. Knowing which plants, grains, and seeds were being moved, and when, gives an idea of which countries and cities were communicating and trading at certain times. This can be linked with knowledge on whether cattle were present to trade at the same time. In these ways, the sciences, both animal and botanical, and the visual and written arts, help us to track where cattle evolved,

ANCIENT ALLIANCES

Ancient Egyptian murals depicting the domesticated Sanga cattle show that they were important animals in that region. Likewise, the Romans and Vikings portrayed and even wrote about early modern cattle, highlighting their importance in farming. Archeological remains have shown us that cattle and humans lived closely together. In some regions where the weather was clement, the animals thrived in the fields; in other colder regions, they lived in barns and even within the home. We know that cattle were transported and traded throughout regions, which is how they conquered the globe.

existed, and moved throughout time, and how the earliest of relationships with humans were forged. It is likely that there is still much more evidence to be uncovered, which makes for a fascinating history, with potentially more exciting discoveries to come.

Top left: Ancient remains and artifacts, such as this indicine statue, have helped map the history of bovines.

CHAPTER 2

Anatomy & Biology

The Cow as a Mammal ❧

There are distinctive characteristics that are used to define mammals. First, mammals must be vertebrates, with a backbone, which obviously cattle have. A second mammalian characteristic is having a part of the brain called the neocortex, which is involved with language, spatial reasoning, senses, understanding, recognition, sleep, memory, learning, and movement commands. The average brain weight for cattle weighing 1,320 lb (600 kg) is around 1 lb (480 g), and the bovine neocortex is less developed than it is in people or dolphins.

Mammals have hair or fur, and this has an important role for cattle: in cold weather, hair growth is stimulated in order to protect the animal. Adequate nutrition is required to ensure hair growth and condition, both of which vary across the cattle breed types. Scottish Highland cattle have long hair and are, therefore, well adapted to colder temperatures, whereas Senepol thrive in tropical countries, as their short sleek hair helps them better tolerate heat.

Below: *These Scottish Highland cattle certainly show the mammalian feature of hair to its optimum. This breed is well adapted for colder climates.*

Mammals also have three middle-ear bones called malleus, incus, and stapes, collectively known as the ossicles. These bones evolved from jaw bones to become part of the ear, enabling mammals to have better hearing, with greater accuracy and efficiency. The single large bone in the lower jaw is thought to be stronger and more stable, especially when biting, than a jaw containing multiple bones, which is probably why this feature continued in mammals.

Metabolic rates tell us how much energy an animal is using over time. A high metabolic rate is not a defining feature of mammals, as birds also tend to have a high rate, often higher; reptiles tend to have lower metabolic rates. Mammals and birds are endotherms, using their metabolism to maintain a steady, high body temperature; reptiles are ectotherms, which will also gain heat from their environment, such as from the sun.

WEIGHT VS. METABOLIC RATE

On average, animals that move more have higher metabolic rates, and the smaller mass an animal has, the higher the metabolic rate will be.

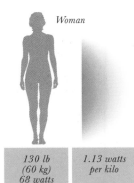

Woman

| 130 lb (60 kg) 68 watts | 1.13 watts per kilo |

Cow

| 880 lb (400 kg) 266 watts | 0.67 watts per kilo |

Steer

| 1,500 lb (680 kg) 411 watts | 0.60 watts per kilo |

Above: *This bull has the single bone for the lower jaw, ideal for eating, and the three ear bones, which result in better hearing—all features of mammals.*

Left: *The woman produces more watts for her weight when compared to the cow, and the cow more than the steer, which backs up the theory that the smaller the mammal, the higher its metabolic rate.*

MILKING IT

The namesake characteristic of mammals is that they produce milk for offspring. Although it is the females that do this, many males have the remnants of the anatomical features but cannot lactate and do not have teats. The word "mammal" actually comes from the Latin word *mamma*, which means breast. Believe it or not, the mammary glands in all species are actually modified sweat glands. The cow has developed an unrivaled capacity over the years for producing milk, and it is these glands that have allowed that to happen. Even more interesting is the fact that mammary tissue development begins in utero during embryonic development, well before milk is required for offspring. In fact, the developing embryo has the tissue that later becomes the mammary glands. The cow develops four mammary glands but this can differ in other species, such as humans and horses, which have two, and other mammals, such as mice and dogs, developing a ridge along which several mammary glands eventually develop in turn. As the cow grows older, the glands develop, but the really large development period is during first pregnancy and lactation. Thereafter, the glands undergo biological and anatomical changes when the cow stops producing milk, and even more changes when she enters later lactation periods. A first lactating cow produces around 70 percent of what a mature cow will in second/third lactations.

Cattle are one of around 5,450 mammalian species. With so much variation in mammals, ranging from carnivores to herbivores, bats to giraffes, predators to prey, it is good to know that cattle are highly recognizable. It is also fascinating to contemplate that despite being from the same species, there is considerable variation between the species and subspecies.

Right: *This Hereford calf drinks the milk from one of four modified sweat glands called the udder and teat in cows.*

Life Cycle

The life cycle of cattle can be split into a number of stages, throughout which their water, food, and nutritional requirements, and medical and housing needs, change, as do their social interactions and temperaments.

Stages 1 & 2: Newborn and calf

A newly born calf is often referred to as a neonate. The first few hours and days after birth are critical. Checking the newborn for abnormalities such as genetic conditions, infections, or disease is vital. In most farming systems the calves stay with their mothers initially in order to receive the vital colostrum milk which contains antibodies. This helps give the calf immunity, protecting it from disease. The bovine placenta is difficult for immunoglobulins to pass through due to all of its layers, in comparison to human placentas. These immunoglobulins pass from mother to fetus, providing immunity, and are also in colostrum, meaning that the early milk is even more important in cows. In many dairy systems the calves will then be fed on bottled milk after a few days, whereas on beef farms the calves often suckle for 4–6 months. On beef farms, most males are castrated (called steers) at a few weeks old, as the hormones produced during puberty give the meat a taste that many people do not like. Males on dairy farms are often slaughtered around birth; some are sold as veal meat or go to feedlots, to be finished out for the meat industry. Castration is often carried out using Burdizzo, Elastrator bands, or surgery, and should be done before the calf is weaned, usually within 2–3 days. In both the beef and dairy systems some males are kept back as breeding stock, but these numbers are decreasing due to the popularity of artificial insemination.

Below: *These calves on a ranch in Utah are being fed bottled milk. Colostrum from the mother is advised for the first few days.*

Stages 3 & 4: Puberty and reproductive ages

A cow starts puberty at 12–24 months old, depending on the breed, environment, genetics, food availability, and a number of other factors. For example, Holstein heifers may not reach puberty until 72 weeks old if fed a low nutrition diet, but can become fertile at just 37 weeks (9–10 months) if on a good diet. The female's ability to conceive, remain pregnant, produce strong and healthy offspring, produce milk, and conceive again quickly are essential for producer profitability and the cow having a productive life. The emphasis on each of these stages differs for the dairy and beef industry, and may not matter so much in smallholdings or for show and working animals.

Many cattle are not used for breeding, but reaching 1–2 years old is still highly relevant. In the beef industry, the goal is to get calves to an ideal weight, at which point they are called "finishers." This frequently happens at around 18 months old, when they are slaughtered and sold to the meat and associated industries, with a few kept back for breeding purposes. Bulls start producing sperm continuously upon reaching puberty, and although the numbers and quality may reduce over time and due to other factors such as temperature and illness, technically, they are able to breed into old age.

Stage 5: Old age

Cattle can live for over 20 years. It is fair to say that most cattle used in the beef, dairy, and related industries do not reach natural old age. Typically, dairy cows live for 4–10 years, 5 on average in high-yield developed countries, and 9–10 in less developed areas. Once fertility drops, lameness sets in, milk production declines, and meat quality reduces, meaning the cattle are no longer profitable. Females aged 5–6 years old who have had 4–5 lactations usually have lower milk yields.

Left: *This elderly African buffalo is grazing in the Maasai Mara National Reserve, Kenya. Buffalo, bison, and domesticated cattle have life expectancies of around 20–25 years.*

LIFE EXPECTANCY

Big Bertha, an Irish cow, died just shy of her 49th birthday and calved 39 times, setting both age and calving records. She was somewhat of a celebrity, and even helped raise money for cancer charities. Hamish the Highland Bull in the UK lived until he was 22 years old; older bulls have been recorded, especially in India.

The Reproductive System ✎

Successful breeding in dairy and beef cattle is essential for many owners. Generally, cattle, especially females, must be fertile in order to be economically viable. In rare breeds, reduced fertility is a worry as the genetic line may not continue.

The female

The ova (female eggs) develop inside two ovaries. The eggs develop from primary follicles which are present before birth; there can be hundreds of thousands of them in every female calf born. Females are born with a finite number of eggs; in contrast, males can continuously produce sperm. The primary follicles must develop into secondary follicles, usually a few will try to develop in each estrous cycle, but usually only one egg is released from the ovary.

The oviducts are tubes that take the eggs from the ovaries into the uterus. The eggs are fertilized within these tubes, and once the egg and sperm combine, the zygote (fertilized ovum) is produced. The zygote is delivered into one of the two horns of the uterus, where it will continue to grow and develop. A layer called the endometrium lines the uterus and undergoes vital changes throughout the reproductive cycle and pregnancy.

The uterus also contains caruncles, structures that enable fetal membranes to attach and the placenta to develop. The uterus thickens, providing a blood supply for the embryo while also physically protecting offspring.

The male

In order for a successful embryo to form, the male must produce active and healthy spermatozoa (sperm), which are produced within the adult testes. Two testes form inside the body, near the kidneys, when the male is still an embryo. Once the male

Below: Bulls become fertile at puberty, at roughly 1 year old with a scrotal circumference of 10 in. (26 cm). Many bulls remain in service until 10–12 years old.

COW'S REPRODUCTIVE SYSTEM

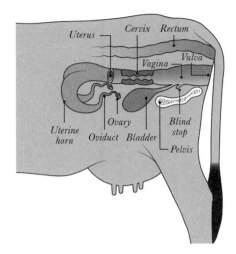

Left: *The egg (ova) develops inside the ovaries and is released into the oviducts, where it can become fertilized. The fetus will develop within a uterine horn.*

Right: *Sperm are created and develop within the testes once a bull reaches puberty. They travel along the ductus deferens and urethra, then are ejaculated.*

BULL'S REPRODUCTIVE SYSTEM

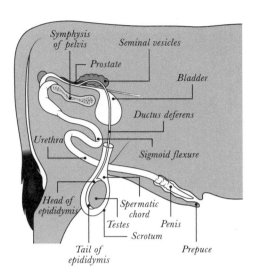

is born, the testes descend in order to be a few degrees lower than body temperature, which is better for sperm production. The testes of the bull are S-shaped and are housed in the scrotum, which helps to protect these essential organs while keeping them the right temperature. The muscles within can contract and pull the testes toward the body when it is cold. If one or both of the testes do not descend (a condition called cryptorchid testis), the bull is likely to be sterile, especially if both testes remain in the abdomen as sperm are not produced. This condition can be inherited, so owners might not breed from a bull even if only one testis is affected. Castration, the removal of the testes, may also be chosen for any bull in order to enhance meat quality and reduce aggressive or mating behavior.

In addition to the testes, a number of other structures help the sperm reach the ovum. Once the sperm leave the testes they travel along the epididymis, spending 10–15 days in this structure, becoming mature in preparation for fertilization. Next, they reach a structure called the ductus deferens, which pushes the sperm into a tube called the urethra, which travels through the penis. A vasectomy is an operation whereby the ductus deferens is cut in order to prevent sperm delivery. These males may be used to help identify females in heat, as the male still produces sex hormones, unlike castrated males. Another important structure includes the accessory glands, which produce semen. When more blood enters the penis through the arteries, and the sigmoid flexure extends, the penis becomes ready to deliver the sperm into the female's vagina, or into one of the artificial insemination collecting devices.

The Biology of Breeding ❧

Cows follow an average 21–22 day estrous cycle (can be 18–24 days), slightly shortened to 20 days in heifers. This includes four main stages: stage 1, proestrous (days 18–20 in the cow); stage 2, estrous (day 21); stage 3, metestrus (days 1–5); and stage 4, diestrus (days 6–17). Important hormones including progesterone, estrogen, follicle-stimulating hormone, and luteinizing hormone will fluctuate in order to mature and release a follicle, and develop and mature the corpus luteum. In the cow, the egg (called an oocyte at this stage) is released regardless of whether or not she has mated. This is particularly useful in modern farming, as artificial insemination can be used.

THE BREEDING REVOLUTION

Artificial insemination has revolutionized cattle breeding industries over the last six decades. Semen is collected from fertile, healthy males, put into semen-extender liquids to preserve the sperm, and stored as frozen samples referred to as straws or units, which can be sent worldwide. A few years ago, India, the country with the largest population of cattle, was producing 66 million straws, covering 20–25 percent of the breedable bovine population. In 2018, around 1.5 million units were sold in the beef industry, 21.9 million in dairy in the USA. The cost associated with both buying the sperm and employing the services of the specialist artificial insemination staff is around $40–60 dollars per live calf born. This differs vastly throughout the world, and sperm prices vary from bull to bull. A prize bull with good quality sperm and a history of high fertilization rates will achieve a higher cost per straw price than other bulls.

Below: *Sperm can be stored in straws and kept cool in specialized tanks containing liquid nitrogen. Research into collection and storage techniques has enhanced artificial insemination semen quality.*

Research has shown that the number and motility of sperm are affected by the collection method, and the number and frequency of sample collection from each animal—even the time of day plays a part. Early morning, when it is quiet and there are few distractions, works with the natural biology of the bull to produce good sperm quality and quantity. Sperm collection is usually carried out using an artificial vagina, by electro-stimulation or by transrectal massage of the accessory sex glands. Much research over the last few decades has concentrated on not only the methods used, but also the methods in which the sperm are stored. Today, artificial insemination is highly successful and commonly used in many countries.

However, there can be disadvantages to this scientific process. It needs specialist equipment, can reduce the breeding gene pool if not carefully monitored, decreases the number of bulls worldwide, requires good knowledge of the female reproductive cycle, and if males are not tested, it can spread disease. Natural breeding also comes with risks: it can spread disease, too; it can reduce the gene pool on particular farms if there is a limited number of males; and females are more likely to get injured. On the plus side, artificial insemination usually increases pregnancy rates; if a sire dies, his remaining sperm can still be used; females that refuse natural methods can often be artificially inseminated; record keeping is often easier; and even older and/or injured males can be used for breeding. There is also an increasing amount of "sexed semen" usage (semen based on the concentration of sperm cells it contains for a particular gender), which is particularly useful for the dairy industry to increase the number of female offspring and reduce the number of males born. It has also enabled beef semen to be used in dairy cows to produce beef dairy crosses.

Below: Artificial insemination can increase pregnancy rates, increase number of male or female offspring, use sperm from bulls around the world, and decrease disease risk.

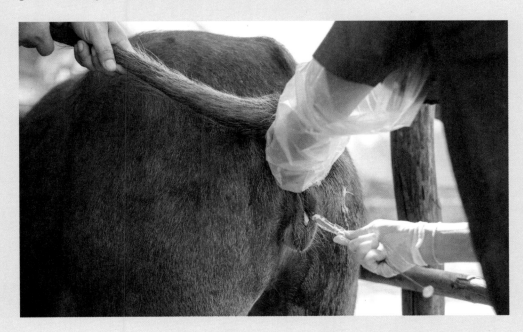

Pregnancy ✍

The timing of mating is vital when trying to achieve pregnancy. Similar to many mammals, for the cow there is a key window of fertility. Whether introducing a bull to a cow or using artificial insemination, getting the timing right is the key to success.

A bull will continuously produce sperm, up to 1.7 billion per milliliter of ejaculated fluid. This decreases if he is mated frequently within a day, and differs greatly depending on the age and breed of the bull. It is certainly higher than human males, who produce between 20–300 million sperm per milliliter. The numbers of sperm are usually more than sufficient for a successful fertilization.

When planning a breeding program, the sperm can be introduced into the female and live for up to 34 hours prior to ovulation, which makes timing fertilization a little easier. However, the sperm must have been in the uterus or oviducts for at least 6 hours in order to fertilize the ovum; in other words, insemination cannot be relied on in those final few hours of estrus and reduced success rates and oocyte quality are seen. Many cows will experience bleeding from the vulva in the few days after estrus, but fertility is generally low during this luteal phase called the metestrus stage.

Below: *There are three stages of calving: 1) cervical dilation (cervix opens); 2) fetal expulsion (calving); 3) placental expulsion (delivery of the placenta).*

IN UTERO

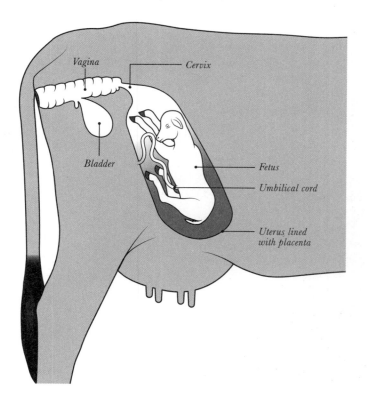

Vagina

Cervix

Bladder

Fetus

Umbilical cord

Uterus lined with placenta

In the female estrus cycle, the latter estrus stage or the metestrus stage is when pregnancy is most likely to occur, but this period only lasts for 15 hours on average, with differences ranging from 6–24 hours. The estrus stage also tends to be shorter in young heifers in comparison to older cows. Although the actual estrus period lasts only 18 hours in cows, sperm can live a relatively long time. Some signs of a cow being in estrous (in heat) include standing to be mounted or mounting other cows, a swollen or reddened vulva, bellowing, and mucus discharge.

FROM ZYGOTE TO FETUS

Pregnancy lasts for around 282 days. It is often detected by feeling within the rectum—called rectal palpation. A blood vessel in the uterus called the uterine artery becomes enlarged during pregnancy and can be felt, as can embryonic membranes and the fetus itself. Once the sperm has fertilized the egg, the newly formed zygote moves along the oviduct and develops in the uterus from days 4–55. The zygote is called an embryo in the early stages of pregnancy and then a fetus in the later stages.

The placenta is a key part of ensuring a successful pregnancy, delivering nutrients, oxygen, hormones, and other essential requirements to the fetus, while removing waste products such as carbon dioxide and food waste. The placenta initiates by day 25 of pregnancy, establishes by day 40, and grows in size as the pregnancy continues. The placenta also produces its own hormones in order to maintain pregnancy and help the fetus and placenta develop. Cattle and human placentas play similar roles and functions but look very different from each other. Human females have one large placenta containing a "fetal" and "maternal" side. The bovine placenta consists of between 70 and 120 placentomes, each measuring 0.4–2.4 in. (5–6 cm) in diameter by the end of pregnancy. The placentomes consist of two parts. The first part is the cotyledon, or fetal, part, and the second is the caruncle, which is maternal. They are linked together by the fusion of maternal and fetal membranes, and appear as buttonlike structures. The caruncles are attached to the mother's uterus and remain inside her after giving birth (parturition), when the caruncles and cotyledons separate and the fetal membranes and cotyledons are expelled from the body within 12 hours. This is often called the after birth.

Below: *Calves are born covered in amniotic fluid, which the mother helps to clean off. Cows eat the delivered placenta to remove the smell of blood, which could be detected by predators.*

The Mechanics of Milk Production ✑

The quantities of milk that a cow can produce are unrivaled by any other mammal. While the biology and anatomy behind milk production in cattle has many similarities to other mammals, at the same time it is a highly specialized process.

Cows have udders, which comprise glands that develop in close proximity to each other. The cow's udder is split into quarters: the right and left half of the udder, which are further split into the front and hind (back) quarter. Each gland (quarter) has several tubes called ducts, which create a system between cells called secretory epithelium, which in turn is supported by fat and connective tissue within a capsule and ligaments. The amount of tissue that secretes milk increases when stimulated by pregnancy and hormones, such as when young are born and when suckling, but then will decrease again when the cow is being dried off. Each gland has a teat, from which the young suckle. In the case of dairy cows, milk is also extracted from the teat during the milking process using a milking machine.

Below: *The udder has four quarters, and although they work independently to produce and release milk, each side shares blood and nerve supplies.*

Although the udder has four quarters, which are separate entities, these do not operate completely independently. The nerves and blood vessels supply each side of the udder. In addition, there are ligaments, called the suspensory apparatus, which support the weight of the udder, and these are also linked between the sides of the udder. This means that the two halves of the udder can work separately. For example, one half of the udder can be removed by a veterinary surgeon in the case of disease, and the remaining half can still function. However, this means that nerve or blood vessel problems can affect the entire side.

DUCT NETWORK

The milk ducts are a complex network, starting with lots of small ducts spreading throughout the tissue, draining into larger ducts, and finally one large duct that leads to the teat. The milk is drained into the teat cistern (just inside the teat) from the gland cistern. The gland cisterns have the capacity to expand greatly as milk gets deposited into them. Surrounding the teat cistern is muscular tissue that prevents the milk from leaving the teat unless suckling occurs. When milk does leave the udder, it goes via the streak canal, a papillary duct, and out through a hole called the ostium papillae at the end of the teat. Although the blood vessels, nerves, and suspensory apparatus are linked between the side of the udder, milk production is carried out in each of the glands separately, independently in the four glands. So while one quarter may stop producing milk or produce less, the other quarters may still produce milk. There are some factors that may stop milk production in all four quarters, too.

Milk production is essential for not only calves but also for humans, and has been for many thousands of years. It has long been regarded as a good source of protein, fats, calcium, and nutrients such as vitamins. Over time, a great deal of research has been carried out into increasing milk yields and quality and understanding diseases and disorders of the lactation system.

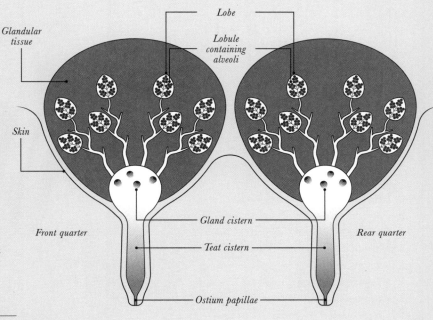

THE UDDER

Lobe

Lobule containing alveoli

Glandular tissue

Skin

Gland cistern

Teat cistern

Front quarter

Rear quarter

Ostium papillae

Above: *The lobules in the udder contain milk-secreting cells called alveolar epithelial cells, the complex system of ducts that transfer milk to the teat.*

Anatomy of a Cow ❧

The anatomy of cattle is often divided into 11 systems:

1. **Integumentary system**: skin, hair, hooves, other related tissues.

2. **Skeletal system**: over 200 bones.

3. **Muscular system**: essential for moving and breathing.

4. **Nervous system**: includes the brain and nerves throughout the whole body.

5. **Circulatory system**: heart and blood vessels.

6. **Lymphatic system**: network of organs, tissues, and vessels helping to remove waste products and fight infections.

7. **Respiratory system**: contains the lungs, ensures cattle breathes in, processes oxygen, and removes carbon dioxide.

8. **Endocrine system**: produces hormones, which help control many daily functions.

9 & 10. **Digestive and urinary/excretory systems**: ensure that food and liquids enter the body, are fully utilized, and are processed before being expelled.

11. **Reproductive system**: acts to produce young.

Right: *Topographical and surface anatomy helps describe regions of the cow based on regions of the body, the shape, and the features that can be seen.*

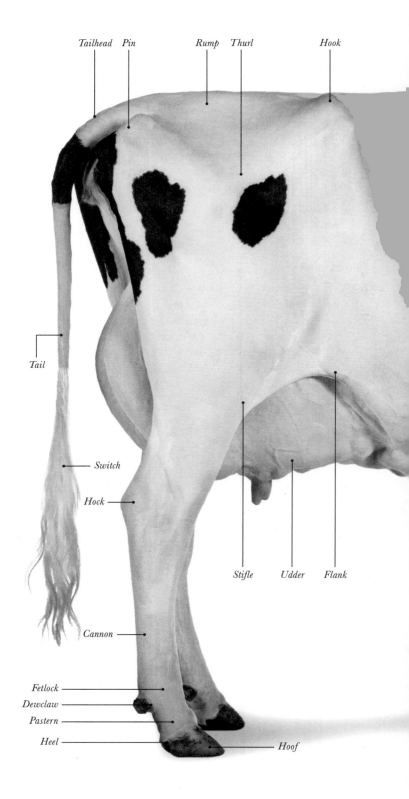

Tailhead Pin   Rump Thurl   Hook

Tail

Switch

Hock

Stifle Udder Flank

Cannon

Fetlock
Dewclaw
Pastern
Heel   Hoof

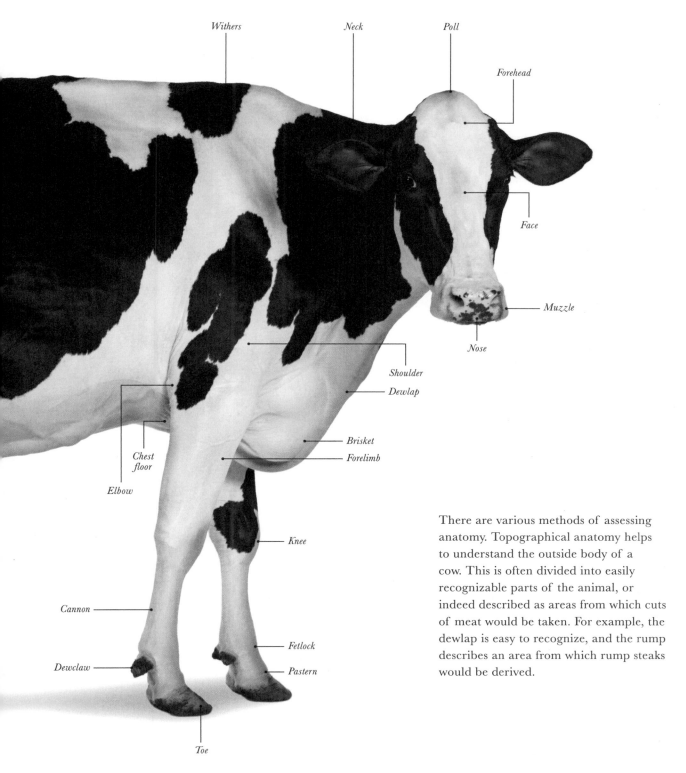

Withers

Neck

Poll

Forehead

Face

Muzzle

Nose

Shoulder

Dewlap

Brisket

Forelimb

Chest
floor

Elbow

Knee

Cannon

Fetlock

Dewclaw

Pastern

Toe

There are various methods of assessing anatomy. Topographical anatomy helps to understand the outside body of a cow. This is often divided into easily recognizable parts of the animal, or indeed described as areas from which cuts of meat would be taken. For example, the dewlap is easy to recognize, and the rump describes an area from which rump steaks would be derived.

The Skeleton ✇

The bovine skeleton is vital in ensuring that the cow can move and stand, and it protects the organs inside the body. Beyond these basic functions the skeleton plays key roles in keeping the animal healthy. Not only does bone marrow produce blood cells, but the bones also help control many factors such as growth hormones, they help to store fat, and they act as a reservoir for vital minerals, including calcium. In the cow, calcium is particularly valuable not only for strong bones, but also in being able to release calcium when needed for milk production. The diet of any cow, but especially one in the later part of pregnancy and during lactation, must be carefully controlled to ensure she takes in enough calcium. Around 7–8 percent of cows annually will suffer from hypocalcemia (low calcium levels, also called milk fever) with the breed, age of cow, lactation stage, nutritional intake, and milk yield all being key factors that affect this number.

BONE DEVELOPMENT

Before bones exist in the growing embryo, cartilage is first laid down, and over time this is replaced by immature bone, which, in turn, develops into mature bone. This process is called ossification, and it does not stop at birth. The growing calf continues to expand its bones. The bones of most young animals are more flexible as they contain more cartilage and less bone—cattle are no exception to this. The larger breeds usually have a longer period of bone growth in comparison to smaller breeds. Each bone will develop at different rates and will depend on factors such as sex, breed, environment, amount of exercise, pregnancy status, and availability of nutrients. In cattle aged 9–30 months old, the lumbar and thoracic vertebrae are not ossified at all. By 42–72 months old, the lumbar vertebrae are usually completely ossified, whereas even at 96 months old, the thoracic vertebrae are still developing.

The hormone estrogen in heifers promotes bone development; therefore, their bones tend to be more advanced in terms of growth than steers, allowing them to mature and "finish" in a more timely manner. Likewise, females who have produced one calf are more likely to have a more mature skeleton than those who have not calved. This effect is so great that estrogen implants are administered to cattle to induce rapid growth, making their carcasses mature quicker for the beef industry.

BONE GROWTH

Bones also develop into slightly different shapes and structures. The skull starts life as many separate bones that maneuver as the fetus is delivered through the birth canal. Over time these bones fuse together, forming a tougher skull. Bones can also get larger if they are used more, for example, the bones of a working ox must grow to support the well-developed muscles. In adult cattle the bones grow in girth not length, but young bovine bones can expand in both directions.

Bone size matters: not only does it indicate whether an animal is growing properly, but also the sizes of the pelvic bones in females can present problems during birth. Bone size, development, and the skeleton in general help us to age cattle.

DAMAGED BONES

Microfractures, fractures, or broken bones require mending throughout the life of cattle. This process is usually carried out by immature bone (woven bone) repairing the main fracture, and mature bone replacing this over time in order to strengthen the structure. There is growing evidence to suggest that microfractures increase the chances of bovine lameness. In cattle, fractures are more likely to occur in animals younger than 12 months old. Fractures of the long bones, such as those in the limbs, are often not treated unless the animal is young. The affected area can be immobilized by splints, casts, and internal and external fixation methods, while walking blocks can alleviate the strain.

Below: *The adult bovine skeleton usually contains 207 bones, similar to adult humans.*

Appendicular skeleton: bones in the limbs.

Axial skeleton: bones in the head, thorax, and vertebrae, including the tail.

Backbone: 49–51 vertebrae forming the backbone—7 cervical vertebrae (neck region), 13 thoracic vertebrae (ribcage), 6 lumbar vertebrae (loin region), 5 sacral vertebrae (sirloin area), and 18–20 caudal vertebrae in the tail region.

Skull: undergoes vast changes during development from a fetus to an adult, and the growth of horns further changes skull anatomy.

Visceral skeleton: bones in soft tissue, including the ossa cordis (heart bones).

BOVINE SKELETON

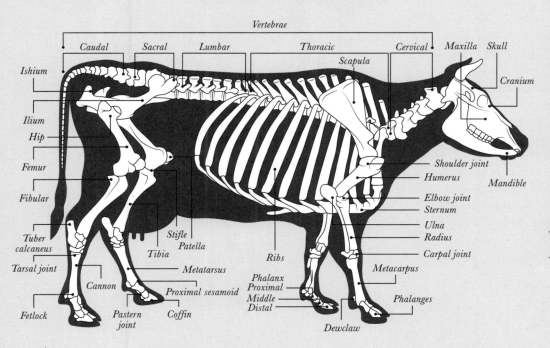

Vertebrae

Caudal — Sacral — Lumbar — Thoracic — Cervical — Maxilla — Skull

Scapula

Ishium

Cranium

Ilium

Hip

Femur

Fibular

Shoulder joint

Humerus

Mandible

Elbow joint

Sternum

Ulna

Radius

Carpal joint

Tuber calcaneus

Stifle

Patella

Tibia

Tarsal joint

Cannon

Metatarsus

Proximal sesamoid

Ribs

Metacarpus

Phalanges

Fetlock

Pastern joint

Coffin

Phalanx
Proximal
Middle
Distal

Dewclaw

The Head & Horns ~

Below: *The skull changes shape when developing from a newborn calf into an adult, and further alterations occur during horn growth.*

The head, skull, and horns of cattle are iconic in many cultures. Some Native Americans use the skull to symbolize life-long protection from the elements, sports teams use it to show strength and power, indigenous Balinese people traditionally carve skulls into works of art so they are not wasted. The main function of the skull, to the cow, is to protect the brain, support sense organs such as the eyes, ears, and nose, and to provide entry points for the digestive and respiratory systems, enabling drinking, eating, and breathing.

The skull is split into two sections: the cranial (back) part, which surrounds the brain; and the facial (front) part. It includes bones such as the occipital, parietal, interparietal, frontal, and temporal bones. In the case of horned cattle, the frontal bones have bony processes that arise from the frontal bones; these can eventually form horns. The facial part of the skull is divided into three basic regions: the orbital, nasal, and oral; or, more simply, the eye, nose, and mouth areas, including the mandible and teeth.

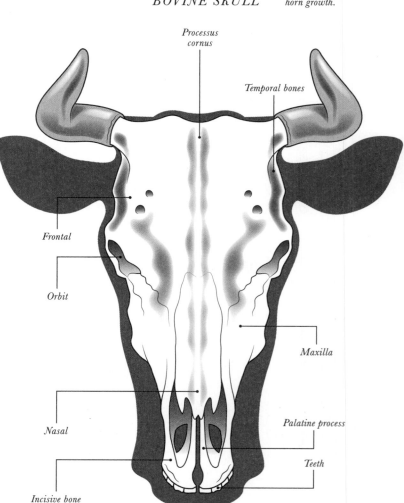

BOVINE SKULL

Processus cornus

Temporal bones

Frontal

Orbit

Maxilla

Nasal

Palatine process

Teeth

Incisive bone

RECORD BREAKERS

The horns contain a central bone called the cornual process, which, by the age of 6 months, is hollow. This structure is surrounded by thickened skin. While many animals have horns, some have antlers, such as deer. Antlers are temporary, for example, during rutting seasons they are shed and another set grow, whereas horns are permanent. Horns have certainly set some fascinating records over the years. The present record holder for the longest horns, measuring 101 in. (257 cm) long, belongs to a Texas Longhorn bull called Cowboy Tuff Chex. This prize-winning bull was sold for $165,000, and each semen straw costs $325. The longest bull horns ever recorded each measured 55 in. (140 cm) and had a circumference of 311 in. (79 cm). Sadly, the Indian Bull named Gopal had to have them surgically removed as they were causing health problems. The longest horn-spread record goes to Poncho Via, a Texas Longhorn from Alabama, at 10 ft 7 in. (3.24 m), twice the width of a concert grand piano. At just 7 years old, there is still time for his horns to grow even longer. However, even with aging it is unlikely that he will reach the all-time longest horn record of any animal. This credit goes to an Asian water buffalo (*Bubalus arnee*) living in India, who had a staggering span of 13 ft 10 in. (4.22 m) in 1955.

An Ankole-Watusi steer named Lurch had the largest horn circumference at 37 ½ in. (95.25 cm); unfortunately, he died of cancer that originated in the horns. The bull with the largest circumference was CT Woodie, also an Ankole-Watusi measuring in at 40¾ in. (103.5 cm). Due to their hormone and growth levels, horn comparisons are made separately between bulls and steers.

HORNLESS?

There is a common misconception that females do not have horns, but this is not true. Their horns do tend to be smaller than those of males, but whether horns are present in an individual is a matter of breed and genetics rather than sex. Some cattle are naturally hornless (called polled), while some have their horns removed (dehorned) or prevented from growing (disbudded), often by veterinary professionals. Cattle are normally dehorned or disbudded for safety reasons, as horns can cause harm for other animals, and the people working with them. At other times, cattle may be dehorned or disbudded to assist the animal with any health-related issues they may cause—for example, if the horns get too large, break, become cancerous, or cause infections or other complications. Disbudding is more usual as dehorning can be painful for the animal. Surgery is more usual as the animal gets older, but often the buds are cauterized or a paste is applied which prevents further growth.

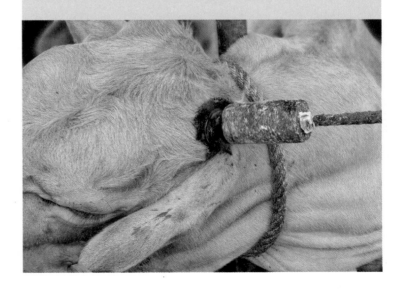

The Teeth & Jaw ❧

The bovine adult jaw contains four types of teeth: incisors, canines, premolars, and molars. In total there are 32 teeth, consisting of six premolars and six molars on the upper row, and six incisors, two canines, six premolars, and six molars on the bottom. The milk teeth in calves look quite different, consisting of six deciduous premolars on both the bottom and top of the mouth, two deciduous canines on the bottom, and six deciduous incisors on the bottom of the jaw. The upper incisors and canines are missing in cattle and, indeed, most domestic ruminants. These are replaced by the dental pad, which is connective tissue with a strong cornified epithelium. Although these teeth are absent in the adult, bovine embryos actually have traces of these teeth within their jaws.

Teeth in bovines have a similar structure to other mammals, with a crown above the gum, a root within the gum, and the neck region connecting the two. The crown is covered in enamel with layers called cement, dentin, and then pulp cavity. The tooth structure is a method by which animals can be aged. The permanent first two sets of three incisors erupt at different ages: at 1.5–2 years for the incisor called I1, then 2–2.5 years for I2; 3 years for I3; and

finally the two canines erupt at 3.5–4 years old. Cattle are also different from carnivorous animals in general, as the incisors and canine teeth have a similar shape; some suggest that the canines are really the fourth set of incisors in cattle. These incisors and canines, termed brachydont teeth, grow fully and then erupt from the gums. In contrast, the premolars and molars (often collectively called cheek teeth) are hypsodont and continue to grow longer after they have erupted. As the teeth age, they grind down over time and become less sharp.

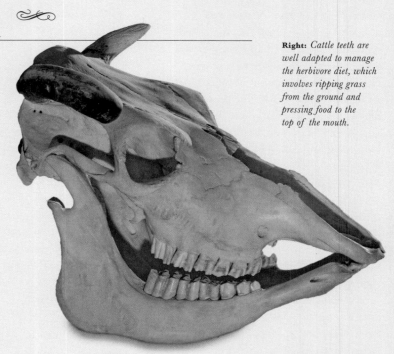

Right: *Cattle teeth are well adapted to manage the herbivore diet, which involves ripping grass from the ground and pressing food to the top of the mouth.*

Above: *Teeth are especially useful for aging cattle as the specific types of teeth grow and grind down at different rates.*

The lower jaw (mandible) of cattle is narrow in comparison to the top jaw. The chewing motion of cattle is lateral (side to side), which means that the teeth get worn down at a similar rate as chewing happens on one side of the jaw and then the other. When eating, cattle press the grass/food toward the top of their mouth (the hard palate) with their lower teeth. The hard palate replaces the top teeth and is present in many herbivores such as deer, sheep, and giraffes. This is a very different technique in comparison to carnivores and omnivores such as dogs and humans. It also explains why cattle do not really bite people, although they can "gum" you instead.

Right: *Each tooth has a crown, neck, and root. Cattle have incisors, canine teeth, premolars, and molars. The canine teeth at the front are incisor shaped.*

BOVINE TEETH

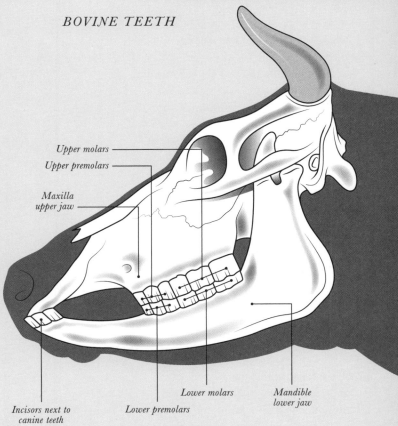

Upper molars

Upper premolars

Maxilla
upper jaw

Incisors next to
canine teeth

Lower premolars

Lower molars

Mandible
lower jaw

CHEWING THE CUD

The teeth, jaw, tongue, and muscles on the bovine head such as the masseter and temporalis enable cattle to grasp then chew food. The muscles in the bovine jaw and head differ in comparison to carnivores, enabling them to grind their food rather than bite down and chew, as required for predators. The temporalis is larger in carnivores and is a strong muscle spanning from the molars to the temples and back to the ear. In contrast, bovids have an extremely large masseter muscle in the cheek area to assist with chewing. The medial and lateral pterygoid muscles also assist with the sideways action of the jaws. They are deeper into the skull than the masseter. There are three main glands—the parotid, submandibular, and sublingual glands—which produce saliva and enable cattle to eat dry food, and enzymes that help the process of breaking the food down.

As the mouth is such an important part of day to day life for cattle, those born with severe congenital problems such as cleft palate are usually euthanized to ensure welfare of the animals. If milk is present in the nostrils when suckling, it indicates to the farmer that the palate should be checked. Like most other animals, the teeth, jaw, and soft tissue can get broken and infected. Conditions such as lumpy jaw, wooden tongue, and calf diphtheria are all bacterial infections that require antibiotics. Lumpy jaw, actinomycosis, is an infection caught when permanent molar teeth erupt or there is injury to the area. The jaw will enlarge; the condition is very painful and will reduce

MUSCLES AND GLANDS OF MASTICATION

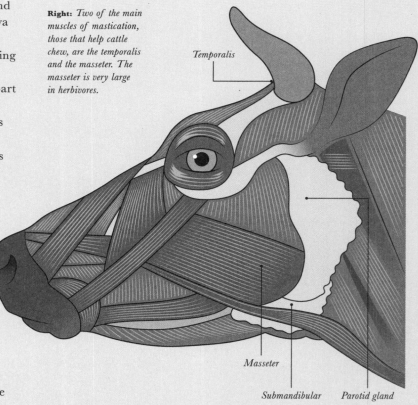

Right: *Two of the main muscles of mastication, those that help cattle chew, are the temporalis and the masseter. The masseter is very large in herbivores.*

Temporalis

Masseter

Submandibular   *Parotid gland*

Parietal crest

Above left: *Chewing the cud is essential in cattle, not only to break down food physically but also by using enzymes present in the saliva.*

Above: *It is rare for cattle to recover completely from Actinomycosis of the maxilla, also called lumpy jaw. This infection is caused when teeth erupt or as a result of injury.*

Calf diphtheria symptoms might include swelling of the cheek and excessive drool being produced. If left untreated, the infection can spread and ultimately cause death.

Jaw fractures are uncommon, as the mandible is relatively strong, but they can be caused by collisions with farm machinery. During birth, the jaw can be fractured when trying to reposition the head of the calf during a difficult birth or malpresentation. Jaw fractures may be treated by confining the animal and feeding it soft foods alongside administration of antibiotics, and complex surgical procedures are available to support broken jaws; but severe fractures usually result in euthanasia.

Although most owners do not brush cattle teeth, oral hygiene and care concentrates on checking tooth and mouth condition for infection. Cattle do not eat refined sugars, which cause tooth decay, and they do eat fibrous food, which cleans the teeth, like a natural toothbrush.

milk yields, as the cow will have problems eating fibrous foods. Although treated with antibiotics, most animals with lumpy jaw are sent for slaughter, as the infection will likely reoccur or go into temporary remission. Wooden tongue infections result in painful, swollen tongues.

STRANGE BUT TRUE

- President George Washington had dentures made from cow, hippopotamus, and walrus teeth.

- Cattle teeth and skulls are sold across the world for decorative effect and making jewelry.

- Some toothpaste is made from bovine products such as dung, ghee, and urine, and often contains milk.

- Cow teeth can be boiled in water and the resulting liquid used for medicinal purposes.

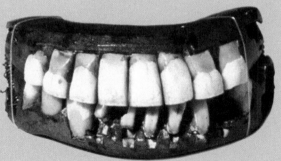

George Washington's dentures

Eating & Digestive System

The reason for having a digestive system is that the body cannot directly use the food consumed. Food must be broken down into amino acids, fatty acids, and monosaccharides, so that each cell can make use of the carbohydrates, proteins, and lipids in the food. The mouth holds food and produces saliva in order to start breaking down the food using enzymes. It's here that the digestive system in bovines is a little different from other mammals. Cattle have an enzyme called lysozyme and salivary amylase, which is present in pigs and horses, although, most ruminants do not produce amylase. In order to deal with their dry, fibrous food, cattle need to produce around 420 pints (200 liters) of saliva a day, in comparison to the 2–4 pints (1–2 liters) produced by humans.

Below: *Cattle often eat dry food products such as hay, therefore their salivary glands must produce hundreds of pints of saliva every day.*

In addition to chemically breaking down food using saliva, the very act of chewing also serves to break down the food ensuring a larger surface area for enzymes to work on. The lips of cattle are not as soft as observed in other mammals such as the horse, therefore they are less flexible, but they are useful for gripping on to food, as is the tongue. The tongue is very muscular, which enables it to move in a number of directions easily, thereby grasping and manipulating food effectively. The roughness of the tongue comes from little outgrowths called papillae. There are several types in cattle called filiform, fungiform, vallate, and conical. The conical papillae are only

Right: *The tongue is not only useful for gripping food, it also contains papillae, which are often called taste buds. They make the tongue feel rough.*

Below: *Cow saliva contains the enzymes lysozyme and salivary amaylase. After chewing, the food is swallowed, but returns to the mouth when the cow ruminates.*

present in ruminants such as cattle, and, importantly, the foliate papillae are mostly absent in cattle (they are found in some individuals) whereas they are present in horses, dogs, and pigs. The filiform and conical papillae do not contain taste buds but the other types do.

Once the food has gone into the mouth, it must pass through the pharynx. This tube contains muscles that ensure that the food and water pass down the esophagus while air is directed into the ventral larynx. Also in this area, cattle, like humans, have tonsils. There are four sets called the pharyngeal, tubal, lingual, and tonsil of the soft palate. The palatine tonsils are the ones that we associate with tonsillitis, and these are in a similar place in the mouth in cattle. The tonsils are often associated with immunity and fighting infections. Cattle tonsils can become infected, in which case they should not make it into the food chain or be consumed by humans.

INTO THE STOMACH...
AND BEYOND

Once swallowed, food must travel down the esophagus to the stomach. The ruminant stomach is remarkable. Although many people believe cattle have four stomachs, it is actually one stomach with different regions. Most mammals have four distinct parts of the stomach called the esophageal, cardiac gland, fundic gland, and pyloric gland. In cattle, the esophageal region has three parts to it, called the reticulum (honeycomb), rumen (paunch), and the omasum (also known as "manyplies" or "the bible," due to its many leaflets). These are called the forestomach. Sharp or metal objects swallowed by cattle often get caught in the reticulum, and can remain there or pierce the stomach lining. The reticulum and rumen together are extremely large and can hold 30–60 gallons/110–235 liters in an adult. Food enters the forestomach, is digested by bacteria, and then passes through into the gland regions of the stomach: the cardiac, fundic, and pyloric, which are also called the abomasum or the true stomach. In calves, milk bypasses the rumen and directly enters the abomasum.

Bacteria are vital in cattle. Mammals cannot break down plant food directly, as they do not produce an enzyme called cellulose. This is a problem for cattle as their diets are plant-based. Therefore, bacteria and other microorganisms live in the digestive system and produce cellulase; this process is called fermentative digestion. Bacteria also produce B vitamins, vitamin K, and amino acids, and are a source of protein themselves. This process of fermentative digestion (fermentation that is a fermentative digestion process) produces around 30–50 quarts (28–47 liters) of gas per hour, including methane and carbon dioxide. Cattle belch in order to remove these gases, as build-up can result in a condition called bloat. Bloat increases the pressure within the animal and can cause heart and lung function to cease— ultimately causing death. It is often caused by eating too many legume plants. With early symptoms including a distended left abdomen, a reluctance to graze or move, rapid breathing, and distress, the animal can be treated with anti-bloat medication. Avoiding pastures with legumes growing such as clover can help avoid the danger, as can increasing fiber intake by feeding the cattle hay.

After the stomach, food travels into the small intestine, which itself is split into three parts: the duodenum, jejunum, and the ileum. The pancreas produces sodium bicarbonate and digestive enzymes which are released into the duodenum, assisting with breaking down the food. The contents of the small intestine then enter into the large intestine, which consists of the cecum, ascending colon, transverse colon, descending colon, rectum, and anal canal. The main role of the intestine is to absorb nutrients into the body via blood vessels, to absorb any water into the bloodstream, and, of course, to release waste products. The intestine of a cow is roughly 20 times the length of the individual, around 130 ft (40 m) on average. This is comparatively much longer than other animals. A whale intestine can be 500 ft (150 m), but this is only eight times the length of its body. Generally, in herbivores the small intestines can be 25–75 times its body length, whereas in carnivores it is more likely to be 4–8 times the length. It also varies in mature beef cattle, at 93–140 ft (28–43 m), compared to dairy cows, ranging from 144–172 ft (44–52 m) long. The large intestine is actually shorter than the small intestine, ranging from 23–41 ft (7–12 m) in beef animals to 43–46 ft (13–14 m) in dairy cows.

The liver is also an essential organ in bovines. It has a key role in metabolic functions, in addition to being part of the immune system. The liver detoxifies and eliminates waste, controls blood glucose levels, and helps to control immune responses, among many other functions. It produces a secretion called bile, which is stored and concentrated in the gallbladder before being released into the

duodenum, to aid with digestion of lipids. Liver fluke (*Fasciola hepatica*) infections are common, as grazing cattle can pick up the parasitic larvae which migrates into the liver. It can have a chronic effect on their body weight and condition, may cause anemia (shortage of hemoglobin and red blood cells), and can make cattle more susceptible to other infections. It also costs the industry in economic terms, as it reduces milk yield and quality. With the parasites becoming resistant to the drugs (anthelmintic resistance), along with changing climate, farm practices, and an increase in cattle transportation and movement from pasture to pasture or even differing countries, the incidence of parasitic infections is steadily increasing.

Below: *It is hard to believe that 130 ft (40 m) of intestine fit within a cow. The digestive system is essential for breaking down and absorbing vital nutrients.*

THE DIGESTIVE ORGANS

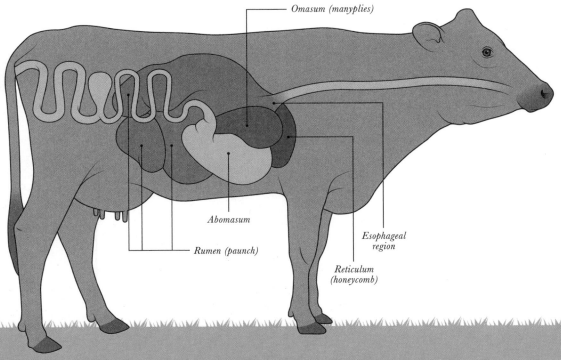

Omasum (manyplies)

Abomasum

Rumen (paunch)

Esophageal region

Reticulum (honeycomb)

How the Hoof Developed ❧

Cattle are even-toed ungulates: they bear weight on an even number of toes, in this case, two. Their hooves are also described as cloven, and sometimes described as a cleft, divided, or split, which means the hoof is split into two toes. The hooves are very specialized in cattle and have evolved considerably over the years in order to bear the bodyweight and enable them to walk in differing terrains. Cattle have two elongated middle digits in each hoof, plus rudimentary dewclaws, which don't bare any weight. In contrast, humans and mice have five digits, horses just one, and the rhinoceros has three. Recent scientific studies in bovine genetics show that several genes play a large role in creating two digits from the original five. Geneticists and paleontologists believe that this change probably happened around 55 million years ago, in the Eocene period, well before the known ancestors of modern cattle.

The hoof protects the skeleton, soft tissues, blood vessels, nerves, and other tissues within the structure. Embryonic cattle show the developing digits by just day 32 of pregnancy. The hoof itself is made from modified skin, with a very thick outer layer called the epidermis. The epidermis has no blood vessels in it, but the underlying dermis layer does contain blood vessels. This dermis layer in the hoof is usually called the corium.

The waxy outermost layer of the hoof is called the periople; the coronary band is the point at which the haired shin becomes hoof; the hoof wall grows from this epidermal skin. Between the hoof wall and the corium lies tissue called laminae. The horny sole and bulb of the hoof are in contact with the ground, and above the bulb lies a fatty tissue called the digital cushion, which helps to absorb shock when the animal is standing or moving.

Below: *Around 55 million years ago, five digits had evolved into just two in the early ancestors of cattle.*

HOOF PARTS

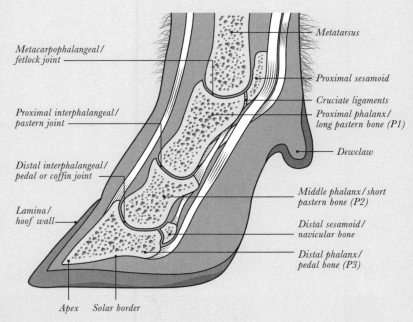

Metacarpophalangeal/ fetlock joint

Proximal interphalangeal/ pastern joint

Distal interphalangeal/ pedal or coffin joint

Lamina/ hoof wall

Apex   Solar border

Metatarsus

Proximal sesamoid

Cruciate ligaments

Proximal phalanx/ long pastern bone (P1)

Dewclaw

Middle phalanx/short pastern bone (P2)

Distal sesamoid/ navicular bone

Distal phalanx/ pedal bone (P3)

HOOF HEALTH

In cattle, 88–92 percent of lameness cases involve structures of the foot. Some people may shoe their cattle (an oxshoe), especially if it is used as a draft animal. This process is more difficult than in horses as cattle cannot balance well on three legs, therefore they may need shoeing while lying down in ox slings, which lift the animal up, or shoeing stalls and/or a crush. Trimming is more commonly used to promote foot health, and recent advances in trimming technology have helped guide training, as both under- and over-trimming can have poor health outcomes for cattle.

The two-digit hoof structures comprise several bones, muscles, ligaments, and tendons, in addition to a blood and nervous system. Important bones include the proximal, middle, and distal phalanxes (also called P1/long pastern bone, P2/short pastern bone, and P3/ coffin bone/pedal bone), and the navicular and two sesamoid bones. Ligaments within the hoof have elastic properties and bind bone to bone. Examples are the cruciate ligaments binding the two digits of the foot together.

A number of tendons help foot movement by attaching muscles to the bones. There are three major joints within the foot: the metacarpophalangeal/fetlock joint, proximal interphalangeal/pastern joint, and the distal interphalangeal/ coffin joint, all of which enable the foot to flex and extend.

Above: *The hoof has undergone much adaptation and evolution in comparison to many other species.*

Right: *Hooves are one of the most important areas of cattle as lameness and hoof infections and problems are very common.*

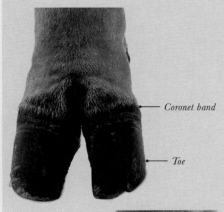

Coronet band

Toe

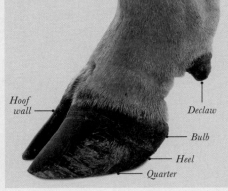

Hoof wall

Declaw

Bulb

Heel

Quarter

Limbs & Lameness &

The names of the bones, muscles, blood vessels, and other anatomical structures of cattle may seem very familiar because, where possible, bones of the mammalian skeleton are named in a similar fashion for all species. The limbs (legs) of cattle are similar to that of the horse, and joints such as the coffin, pastern, and fetlock are named the same in both species.

Cattle have the thoracic (fore or front) and the pelvic (hind or back) limb. On the thoracic limbs there is a scapula bone, humerus, radius, and ulna bones, leading down to the carpel and metacarpal bones, with phalanges forming two strong digits and two dewclaws (small digits), and finally the sesamoid bones. The dewclaws do not reach the ground unless the land is soft, unlike the two digits which are on the ground.

The pelvic (hind) limb starts with the pelvis bone, which in turn articulates with the bones of the pelvis on the main body. Further down the pelvic limb is the femur bone, leading down to the patella (hock area), then the tibia and fibula, which has a relatively small size ratio in comparison to other species, and finally the lateral malleolus. The tarsal bones follow, then the metatarsal bones, phalanges (digits), and sesamoid bones. If you compare the limb bones to other species, many have

similar bones but the big difference is the shape of each bone. The shape has evolved over time to the specific needs of cattle, and, indeed, will change in a young versus an older animal, and even adapts when a lot of muscle is present in particular breeds or individuals.

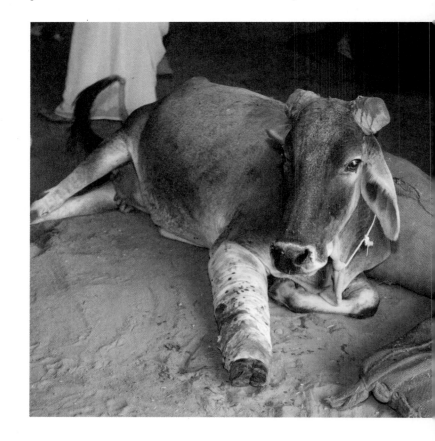

Below: *Most lameness occurs due to disease, but injuries such as fractures can occur, which can often be treated.*

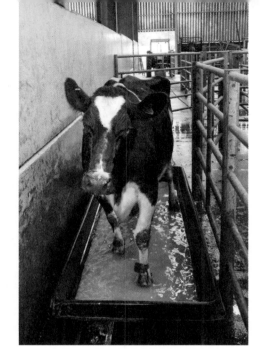

but older cattle show numbers at around 15 percent. Lameness reduces weight-gain abilities, causes pain, and presents problems with transportation, and so is of significant concern for the industry. Consequently, owners, scientists, veterinary professionals, and production industries are making a concerted effort to detect, treat, prevent, and understand lameness, with the ultimate aim to reduce incidence.

Left: *Foot bathing is commonly used to reduce infections and, therefore, lameness. Performed incorrectly it can spread disease and chemicals can burn skin.*

Below: *Knowledge of the limbs helps understanding of lameness, enabling farmers to take the appropriate preventative measures.*

LAMENESS STATISTICS

Around 70 percent of lameness cases are directly attributable to foot/hoof diseases and mechanical injuries such as crushing the foot, or equipment injuries. Injuries to the upper skeleton or major muscles, septic joints, and injection site lesions can also cause lameness.

Recent scientific research showed that extra bone growth was observed and associated with lameness in cattle, a debilitating disorder that is still being researched to further understanding. Body condition scores are used to assess relative fatness and condition of an individual cow. Cows with low body condition scores and a history of lameness are more likely to become lame again, and to produce lower milk yields. Numbers indicate that at any one time around 22–50 percent of dairy cows are lame; this lowers to around 26 percent in bulls. Young beef cattle are rarely lame (under 5 percent)

LIMB PARTS

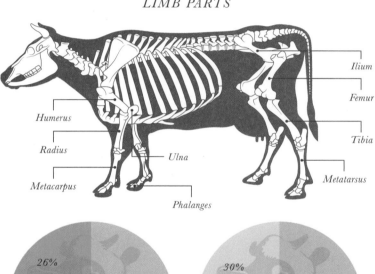

Humerus

Radius

Metacarpus

Ulna

Phalanges

Ilium

Femur

Tibia

Metatarsus

The most common fractures in cattle

- Metacarpus & metatarsus
- Tibia
- Radius and ulna
- Humerus
- Other (inc. femur, pelvis, and phalanges)

The most common causes of lameness in dairy cattle

- Sole ulcer
- Sole hemorrhage
- Digital dermatitis
- White line disease
- Other*

*including physical injury, dog bites, poor nutrition, and genetics

26% 50% 5% 7% 12%

30% 39% 8% 10% 13%

The Heart & Organs ✍

The cardiovascular system, the heart, and blood vessels ensure blood is pumped around the body, delivering nutrients, oxygen, cells, platelets, hormones, and other substances while removing waste products. In order to produce 1 gallon (4 liters) of milk, 400–500 gallons (1,500–1,800 liters) of blood will pass through the udder. The heart structure of cattle is similar to many mammals, including two atria, two ventricles, and valves preventing blood backflow. Similar cardiovascular diseases are observed in cattle as other mammals such as cardiomyopathy, congenital heart defects, and heart failure. On average the bovine adult heart pumps 16,000 gallons (72,700 liters) of blood per day, weighs 2 lb (900 g), is 10½ in. (27 cm) long, and is four times larger than human hearts. Heart rates vary depending on the situation, environment, and individual. A calm cow on a small farm averages 63.5 beats per minute when lying down, rising to 72.4 when milking; a temperamental cow on a large farm would be nearer 81 and 92 beats per minute.

A rare feature of the bovine skeleton is the presence of ossa cordis, or bones in the heart. Often there are two bones present, called the os cordis dexter (heart bone left) and os cordis sinister (heart bone right). The larger heart bone on the right is around 1–2 in. (3–6 cm) in size. The smaller bone more to the left is usually around 0.8 in. (2 cm). These heart bones are thought to help muscle bundles and valves attach to strong tissue, prevent valves from stretching too much, and help electrical signaling pathways throughout the heart. Although a number of bovids have ossa cordis, it is very unusual in other mammals. Given that cattle hearts are eaten throughout the world in popular dishes including soups and steaks, and are grilled on skewers in Peru, care must be taken to remove these bones before consumption.

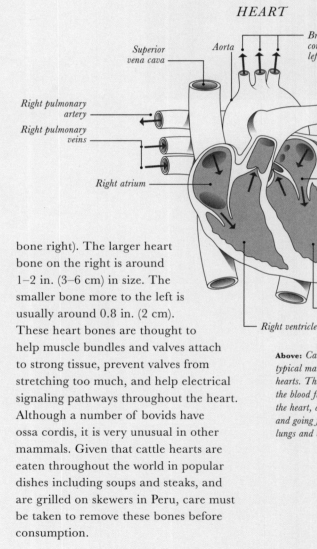

HEART

Superior vena cava

Aorta

Brachiocephalic, left common carotid, and left subclavian arteries

Left pulmonary artery

Right pulmonary artery

Right pulmonary veins

Right atrium

Left pulmonary veins

Left atrium

Left ventricle

Right ventricle

Above: *Cattle have typical mammalian hearts. The arrows show the blood flow through the heart, arriving to and going from the lungs and body.*

LUNGS & KIDNEYS

Although the lungs of many mammals have slightly different shapes, the basic tissue types are similar. The bovine lung is so similar to the human it has been suggested it provides a better model than that of mice. The left lung has two lobes and the right has four. The trachea brings air from the head toward the lungs, and then splits into two tubes, each called a bronchus. Unlike the lung anatomy of most other mammals, the bronchus leading to the right lobe branches directly from the right side of the trachea. Another difference is in lung capacity, the total amount of air which can enter the lungs. Cattle tend to have small capacities in comparison to other mammals such as horses. This means that cattle find lung disorders and infections such as pneumonia more difficult to fight.

The bovine kidneys play a special role in keeping the blood clean, formulating and excreting urine, and maintaining fluid balance within cattle. They must constantly adjust depending on food and water intake, temperature, and the changing needs of the body, while making sure that waste products are excreted. The kidney of most mammals is smooth, but in the fetus it appears as several lobules. The bovine kidney is highly unusual as it stays in the fetal, lobulated shape, appearing to consist of 12 lobules per kidney. Each kidney is also a slightly different shape: the right one is more flattened and ellipsoidal; the left is thicker at one end than the other.

Below: *Key parts of the cardiorespiratory system are the lungs and heart. Kidneys play a vital role in cleaning circulating blood and maintaining fluid balance.*

BOVINE ORGANS

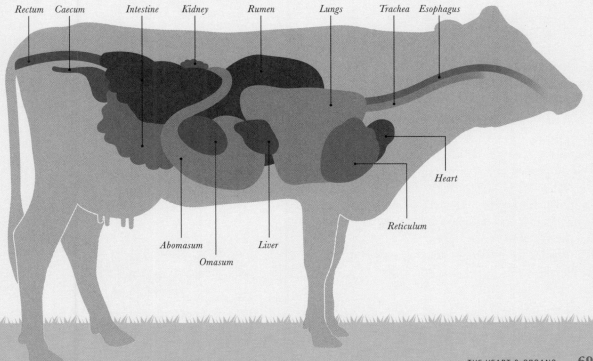

Rectum Caecum Intestine Kidney Rumen Lungs Trachea Esophagus

Heart

Reticulum

Abomasum

Omasum

Liver

Hearing & Sight ❧

Cattle have an acute sense of hearing, better than humans. They can hear in the ranges of 16 to 40,000 Hertz (Hz), whereas in people, this is restricted to 20 to 20,000 Hertz in younger adults. Bovines hear much better in the higher frequencies particularly; it is possible they can even hear ultrasonic noises made by bats. As a prey animal, cattle are constantly aware of their surroundings, and their sense of hearing is particularly important. Additionally, vocalizing with one another is imperative, not only for herd protection but also in discerning needs, especially of calves.

This excellent sense of hearing has been investigated over the years, and it has been found that cattle normally get elevated heart rates when hearing unknown human voices, whether they are calming, low-tone voices or more aggressive in nature. This reaction has been attributed to fear—the animal is preparing for the fight or flight situation. In similar situations, cattle hearing loud banging noises did not react as much, but the overall evidence suggests that maintaining quiet when interacting with cattle is preferable.

The ear is made up of three regions: the external, middle, and inner ear. The external ear contains the visible parts such as the pinna (auricle) and the external part of the auditory canal. Muscles allow the pinna to move in order to detect air pressure and sound; a brown waxy substance called cerumen is produced to protect the canal. Within the middle ear is the air-filled tympanic

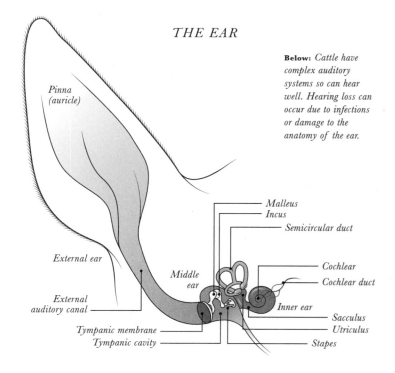

THE EAR

Below: *Cattle have complex auditory systems so can hear well. Hearing loss can occur due to infections or damage to the anatomy of the ear.*

Pinna (auricle)

External ear

External auditory canal

Middle ear

Tympanic membrane

Tympanic cavity

Malleus

Incus

Semicircular duct

Cochlear

Cochlear duct

Inner ear

Sacculus

Utriculus

Stapes

cavity, the three auditory bones (malleus, incus, and stapes, otherwise known as the hammer, anvil, and stirrup). The internal ear detects both sound and acceleration of the head, and contains a system of fluid-filled sacs and ducts, including the utriculus and sacculus, and the semicircular and cochlear ducts. This system is divided into two main parts, the cochlea and the vestibular apparatus. The cochlea is used for hearing and the vestibular apparatus for balance. The main receptor cells within the ear that are responsible for hearing sound are called hair cells. Air-pressure waves are detected using a complex method of vibrations, which are detected by the hair cells after passing through other parts of the ear. This information is then passed into the brain via neurons via the medulla, pons, and midbrain, and into the primary auditory cortex in the cerebrum. Injuries to the vestibular system often result in

a lack of balance, which will mainfest as head tilting, leaning, circling, falling over on one side, or erratic eye movements.

Ear anatomy is important when considering ear tags. Tags are an excellent way of monitoring overall health, production, and ownership, but care must be taken to insert them in the right place, using appropriate materials. If carried out incorrectly, ear health and hearing can be compromised and infection rates can increase. Study results showed that polyurethane tags usually damaged ears less than metal ones. Many people report tagging at specific times of year in order to avoid adverse weather conditions or insect-prevalent times, when cattle are more likely to scratch their ears. Although the ear is comprised of mainly skin and cartilage, making it more pliable than bone, cattle do have a tendency to stick their heads into fences and other places, which can cause physical injury.

Above: *Structures within the ear also help cattle maintain balance. Care should be taken when using ear tags, to reduce infection and damage.*

THE DOMINANT SENSE

Vision is the most dominant sense in cattle. The two eyes sit within the orbits of the skull for protection. Due to the position of the eyes near the sides of the head, cows have 300-degree vision when standing up and higher levels of vision when their head is lowered, in order to feed, for example. This degree of vision is better than humans, who only have 180 degrees as their eyes are more forward facing, but it comes at a price. As cattle cannot see directly behind themselves, they are very cautious of movement behind them, and this area creates the "kick-zone." If you enter this area, the cow/bull has a much higher chance of trying to defend itself, usually by kicking. Cattle also have a blind-spot directly in front of their noses. Due to their eyes being further apart, their binocular vision is also less accurate than that of humans. Therefore, their depth, speed, and distance perception is poor and limited to around 25–50 degrees directly in front of them. The vision that is not binocular is monocular (peripheral) vision, in which the eyes are used separately, and in which useful features such as texture, perspective, motion, and focus are lessened. Cattle also have slit-shaped pupils and relatively weak muscles in their eyes to alter lens shape,

Above right: *With 300-degree vision, cattle have good eyesight. Regular eye checks can detect infections before they escalate.*

Right: *Cattle have compound eyes, with a structure similar to that of humans; however, their color perception differs from ours.*

THE EYE

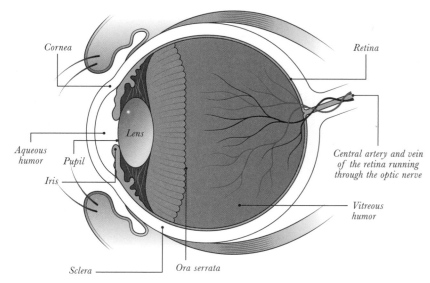

Cornea

Retina

Aqueous humor

Lens

Pupil

Iris

Central artery and vein of the retina running through the optic nerve

Vitreous humor

Sclera

Ora serrata

which means they may focus more slowly in comparison to other species. These traits mean that cows are often wary of shadows, as they may not be sure whether the shadow is a dangerous hole.

Upon receiving light through the pupil, it is focused by the lens onto the retina, which contains photoreceptor cells. Cattle have photoreceptor cells called rods and cones, similar to humans, horses, dogs, and other animals with complex eyes. Rods help the animal to see in all light levels but particularly in darker conditions such as dawn and dusk, whereas the cone cells require more light, as they help detect color. The cones are specific to one of three wavelengths—blue, green, or yellow. In the case of the matador's red cape, the bull is more likely to be seeing the movement rather than the color specifically, although cattle are better at seeing colors with longer wavelengths, including orange and yellow, in comparison to short wavelength colors, such as blue, green, purple, and gray.

Due to the types and numbers of rods and cones, cattle will usually prefer to move away from well-lit or dim areas and shadows. Once the rods and cones receive information, the optic nerve carries impulses along the nervous system to the primary visual cortex within the brain, where information is translated into conscious knowledge. Some of the impulses are used to assess information such as day length—important for seasonal breeders such as cattle.

Right: *Beware of passing directly behind cattle—if they can sense a threat but can't see it, they may kick out their hind legs as a defensive mechanism.*

Smell, Taste, Touch ✃

Like all mammals, cattle rely on their senses for survival. Smell, taste, and touch are vital to ensure that cattle eat appropriate foods, walk on suitable ground, and recognize their herd.

Smell

Olfaction, the sense of smell, relies heavily on olfactory sensory neurons within the nasal cavity. Signals from these cells ultimately transmit into the olfactory bulb within the brain. Olfaction is also known to have an impact on the limbic lobe, a part of the brain generating emotional responses and associated behaviors. A subset of olfactory neurons and a structure called the vomeronasal organ also sense pheromones, chemicals which can influence the behaviors of others. Sex-attractant volatile compounds (pheromones) have been discovered in cattle urine, and various hormones can be present in almost any bodily fluid. This information helps us to understand the act of "flehmen" for reproduction. Flehmen is displayed when bulls and sometimes females raise their noses up and open their mouths slightly in order to detect odors, including pheromones. Cattle can also detect odors such as food, or a cow on heat, from 5–6 miles (8–10 km) away. The smell of blood can

cause fear and panic. In general, cattle have a good sense of smell and rely on it to react to differing situations and assist in key events such as mating.

Taste

With around 25,000–35,000 taste buds on papillae on the tongue, cattle have a good sense of taste, also called gustation. In comparison, the number of taste buds does not rival the catfish, at a significant 175,000, but is superior to chickens, which have just 24 taste buds. Cattle can use this acute sense of gustation by recognizing quickly the bitter tastes of poisonous plants, for example.

Above left: *Cattle have good olfaction: they can smell odors from up to 6 miles (10 km) away. They use this to smell hormones, food, and even blood.*

Above: *In nature, being able to differentiate poisonous food is important; fortunately, cattle have retained their high number of taste buds.*

feed companies also use it to their advantage, by formulating differing flavors in order to encourage food uptake. The four basic taste sensations include sweet, salt, bitter, and sour; however, there is a fifth that detects the amino acid glutamate, which is savory. Silage-based diets frequently have flavoring added to mask the alcohol or vinegar created from yeast fermentation. Some additives are natural, such as garlic, whereas others are fruit extracts or chemicals; older ruminants often prefer citric tastes and aromas. Additives can also reduce the costs of feed in lower socioeconomic areas, as bitter or more rancid, cheap foods can be mixed with flavorings.

Touch

Somatosensation describes the sensations from skin, membranes, limbs, and joints, and is more commonly referred to as a tactile or touch sense. Although cattle have tough skin, they can feel a fly land on them. Pain is also included within the sensation of touch through thermal (hot/cold), chemical, or mechanical means.

The nociceptors (pain receptors) are stimulated by a sensation, and the central nervous system responds automatically to increase mental awareness and elicit behavioral and emotional responses, including escape reactions. The nociceptors are activated when a bull pulls on a nose ring, for example. Normal touch is controlled by touch receptors, which transmit information to the brain, enabling the cattle to respond in a more controlled manner than when feeling pain. Both skin and hair contain mechanoreceptors, which detect touch and skin, and also contain thermoreceptors, to help detect temperatures. The mouth area is particularly sensitive and is often used to explore surroundings using tactile actions. Cows are also more sensitive to electric currents than humans, by a factor of 2:10. Therefore, electric fences are used as a deterrent in certain situations, although use in pre-milking yards as crowd gates is not advised, nor are electric prods, as reactions can result in dangerous behavior.

Below: *Touch is an important sense and people managing cattle also use this to help maintain herds, from electric fences through to nose rings.*

Coat & Color

The ancestors of modern cattle were born a chestnut color before turning a deep brown/black, with a white eel stripe along the spine. They had a light-colored muzzle, some paintings show a light-colored "saddle" on the back and blond, long, curly forehead hairs. Some of the more primitive cattle breeds show many of these features, but breeding has resulted in a vast array of colors and hair and coat types. Two basic pigments, eumelanin (black-brown) and phaeomelanin (red-yellow), and their proportions create most of the colors we see in modern cattle. These two pigments are controlled by a gene coding for the melanocortin receptor 1 (MC1R). In addition, pigment production within a cell is restricted by the abundance of an enzyme called tyrosinase—eumelanin requires higher levels than phaeomelanin.

In some breeds, such as the Charolais and Simmental, a genetic mutation permits the white coat coloration. This is called lightening, or dilution, and creates a range of colors including white, cream, dun, gold, yellow, pale red, gray, and brown. In other species, hundreds of genes have been known to control coat color, which is likely the case in cattle and a few have already been discovered.

Some genes and mutations have been discovered that only appear to affect particular breeds. For example, Dexter cattle have specific genetic coding in the tyrosine-related protein 1 gene, which results in a pale (dun) color, yet other breeds do not have this altered code. The silver (SILV) gene gives Highland cattle their distinct color, but more research needs to be undertaken to see if this is unique.

Below: *Coat color, length, variation, and hair type have long been used to differentiate individuals and breeds, and often enable cattle to adapt to different climates.*

COAT COLORS

Black and white

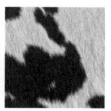
Brown and white

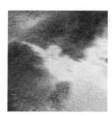
Tan and white

Brown short hair

White short hair

Belted   Dun / gold   Long / wavy hair   Roan

Black   Friesian   Red / brown   White / cream

Whatever the color of the cattle, the genetic code is important, and carries implications for cross-breeding and color variation within offspring. Albino cattle are rare but do exist. True albinism is a lack of pigmentation in the eyes, skin, and / or hair, which is not the same as a white animal. For example, although rare, Highland cattle can be born with white hair despite the usual distinctive rusty red-brown coat. Genetic throwbacks mean that white coats are possible without a lack of pigmentations. Cattle can also have solid color or be spotted. Roan coloration (an even mixture of white and colored hairs) is common in breeds such as the Belgian Blue and Shorthorn, and is controlled, in these two breeds at least, by the mast cell growth factor gene.

Hair lengths and types vary from thick, long, and wavy (for example, Highland cattle and some longhorns), to the slick, shorter hair more typical of cattle living in hotter climates (such as Carora and Boran). The type of coat and color are varied across the breeds, have evolved over time, and often help support the animal in its habitat. Grooming is an essential part of showing prized cattle, and shaving / clipping is a cheap and effective way to keep cattle cooler during warm months, but is also laborious.

Above: *Selective breeding is often used to achieve specific coat types, and research is uncovering the specific genes and mutations responsible.*

Red and white curly

Tan long hair

The Cow Genome ✥

The importance of cattle across the world was highlighted when it became one of the first livestock animals to have its genome fully sequenced and mapped. In 2009 scientists officially unveiled the genome of an 8-year-old Hereford cow from the USA, who made history by becoming the first of her species to have her 22,000 genes made available publicly. Her genes were published in one of the world's leading scientific journals, *Science*, with two original research articles and a special Podcast all published in the same issue. Researchers from 109 universities, institutes, and companies were involved from countries worldwide, including the USA, various European countries, Australia, New Zealand, Brazil, Canada, Japan, and Singapore. The insights both before that stage and afterward have been outstanding, and this understanding of genetic information has helped unravel the genetic diversity, evolution, and disorders of cattle.

Soon after the genome of modern cattle was published, the mitochondrial and whole genome sequences were published in 2010 and 2015 from their ancestors, the aurochs. A bone that was 6,750 years old from the United Kingdom was used to find out the fascinating genetic secrets of the ancestors. This has enabled comparisons of modern cattle and their ancestors, and has helped us to understand the changes that have

Right: *Genetic studies have enabled us to develop knowledge including ancestry and migration throughout the ages, such as information on this nineteenth-century auroch.*

occurred during domestication, including highlighting that the aurochs were the ancestors of both taurine and zebu cattle, and showing that there are distinct differences between the two subspecies. Despite coming from aurochs, genetic analysis also helped to show that these aurochs lived in different locations and were genetically distinct from each other.

TECHNOLOGICAL BREEDING

An example of how genome sequencing has helped is in breed determination— some of this work was carried out even before the whole sequence was published. The origins of some breeds or individuals has not always been obvious but genetic testing can help. A good example is the Dwarf Lulu breed in Nepal, which looks like a taurine but was found to be a hybrid of taurine cattle, indicine (zebu), and the yak. It was also interesting that some of the British traditional cattle had much more of the auroch genome still present in their genome. There are

already many breeding programs that have been designed to reinstate more auroch tendencies. By knowing which cattle have the most similar DNA to these ancestral bovines, it may be possible to target breeding more accurately. These techniques are often called back-breeding and selective breeding.

There are more advanced technologies, such as cloning, available too. The first clone of an extinct animal was the Pyrenean Ibex. The first born clone lived for only seven minutes, but science has been developing and it could become a possibility for the auroch to be cloned, especially as modern cattle could be surrogate mothers for any such clones.

Could we turn back time and reintroduce the now extinct ancestor back into the wild? Could we use genetic technologies to understand present day diseases and disorders in cattle? There is no doubt that advanced technologies are now available that could help us further understand modern cattle, combat disease and illness, and even preserve the species.

CHAPTER 3

Society & Behavior

Housing & Farming ❧

The herd behave differently depending on the type of environment they are in, and have differing needs depending on their ages. If the land is open, with fewer trees and shrubs, the cattle will stand nearer to one another than when there is cover. If the cattle are kept indoors, the calves should be housed separately from older animals, with separate ventilation. Ventilation inlets and windows should be screened in order to reduce flies and insects from entering the unit, to reduce risk. Once weaned, it is still best to group younger, smaller animals in small groups. Most farms aim for 3–5 animals up until 4 months old, increasing to 6–12 animals after 4 months.

A variety of housing options are available, ranging from hutches to shelters and greenhouse barns. Ideally, cattle should be shielded from direct sunlight, kept warm in the winter, be draft free in colder weather, and have adequate waste and urine removal systems.

The air quality and temperature should be monitored—excess dust, for example, can cause respiratory problems. Good ventilation also removes odors, encourages the animals to eat more, removes harmful pathogens, and even helps maintain the buildings themselves. Building design is essential for appropriate ventilation, but fans and sometimes chimneys and a variety of vents are utilized to maintain good airflow; however, if the airflow is too fast it can cause drafts.

If the temperature is too high, cattle will feed less, breathe and sweat more, and become stressed. If the temperature is too low, the animal will use up energy to maintain warmth, which can result in a decrease in growth rate. The optimal temperature for rearing calves is around 60°F (16°C), with 50–85°F (10–29°C) being reasonable; however, the humidity,

Below: *Milk is 87 percent water, therefore lactating cows need 13–22 gallons (60–100 liters) of water a day. Cattle size, breed, and environment also affect water intake.*

wind, and amount of rain and wet mud also play their part in animal comfort.

Optimal humidity for young animals is 65–75 percent. Humidity is not just determined by the moisture in the ambient air and from the weather or environment, such as fields and rivers, but is also affected by breath, urine, feces, and sweat. When determining cow comfort during hotter climates, a Temperature Humidity Index chart (THI) is often referred to in order to help maintain health and comfort. Once temperatures reach 85°F (29°C) and humidity hits 85 percent, beef cattle can be in danger. A temperature of 72°F (22°C) and 100 percent humidity holds the same risk as 85°F (29°C) and 35 percent humidity. A temperature of 72°F (22°C) and 45 percent humidity can expose the cattle to a THI of 68, at which milk losses are often seen. Milk yields reduce, calving rates start to decrease due to lower conception rates, rumination decreases

alongside reduced food intake, rectal temperatures increase, and even disease prevalence—and death—can increase from high temperatures and/or high humidity. During hot periods, it is best to handle animals early in the morning, moving them at slow speeds where necessary. Sprinkler systems can help cool the animals, and shade and good airflow are essential.

Factors such as light, and indeed darkness, help maintain hormone regulation, which can promote quicker weight gain, provide better milk yields, and hasten puberty in heifers. A period of 16–18 hours of daylight followed by 6–8 hours of darkness are optimal for most herds, but facility designs have allowed for the accommodation of these recommended hours without disrupting the cows' ability to maintain comfort. In general, cattle avoid bright light.

Below right: *Calf housing increases welfare and production levels. Calf hutches can be easy to clean, relatively inexpensive, and easy to move.*

THE 5 FREEDOMS

Throughout the world, most laws for welfare are based on the Five Freedoms. Housing and farming standards should also reflect these ideals. The key components for cattle are:

1. Freedom from hunger and thirst—such as access to water and an appropriate diet. Access to the feed will vary in each situation, but space to access food and water troughs is essential, as it reduces bullying and encourages consumption. At least 10 percent of the herd should be able to stand at water troughs at any one time, with each animal drinking at least 4–21 gallons (14–80 liters) per day, depending on whether they are young, suckling, or lactating. The water needs to be fresh and plentiful. Rainwater harvesting is often economical and provides a sound source of water.

2. Freedom from discomfort—which can include good shelter and resting materials. A non-slippery floor with adequate drainage for removal of liquid waste helps create a good environment, as does housing that eases cleaning and provides a positive working environment for staff too. Cattle thrive with the correct bedding material. A cushioned resting surface helps the animals to stay clean; it absorbs moisture, decreases disease risk, and reduces stress. Bedding can range from sand, compost, and straw, but straw does attract the highest amount of flies.

3. Freedom from pain, injury, and disease—which is assisted by providing appropriate shelter, grazing grounds, and healthcare regimes. Adequate ventilation, slip-proof floors, and an absence of sharp objects within housing units and open areas such as fields all help cattle to enjoy this freedom. Floors of indoor housing may also have solid areas to make sure hooves are worn down, thereby reducing lameness. Often, grooved concrete, slatted wood, or rubber are used as flooring surfaces. In the case of slatted flooring, care must be taken that hydrogen sulfide does not build up—it smells like rotten eggs, although at higher concentrations it often emits less odor. The gas is produced from cattle slurry/manure and has even led to the death of some workers. Some farms have automated machinery that scrape the floors, which cattle are trained to avoid when in action, and some even have flush systems that use recycled water to flush the manure out of the freestall barn lanes.

4. Freedom to express normal behavior—often refers to adequate space by not overcrowding animals, and providing companionship and environment enrichment, such as scratching material. A calf kept inside in a loose housing arrangement should have 7–13 ft (2–4 m) squared if it weighs under 330 lb (150 kg), rising to 16 ft (5 m) squared when 330–440 lb (150–200 kg)—this space allowance enables it to groom, stand, lie down, and generally loaf around. Adults need around the same amount of space on a solid floor.

5. Freedom from fear and distress—which is helped by providing a stress-free housing and farming environment, handling by familiar people, and a daily routine. Cubicles are often provided in large sheds for females, especially those housed predominantly inside—males urinate in the center of these, therefore they are not appropriate for male cattle—which can reduce bullying and increase lying time.

Eating & Sleeping

Nutritional needs vary greatly between breeds, ages, gender, and intended use. Whether growing and finishing cattle for the beef industry, ensuring high milk and fertility levels for dairy production, or using cattle for carts and plowing—as a general rule, the larger an animal, the more energy intake is used on maintenance of the existing body, leaving less energy available for growing and milk production. Feed uptake is also affected by diseases such as worms and pneumonia, and conditions such as bullying, temperature, and stress.

When able to graze, dairy cows spend around 8 hours a day eating, and beef cattle around 9 hours, generally grazing to a level of 2 in. (5 cm) above the ground. Grazing is mostly performed in a standing position, whereas rumination is often carried out when resting/not eating. Cattle also regurgitate their food and chew it in order to further break it down, which assists digestion. This action is called "chewing cud," and it indicates that rumination is occurring and that the cow is comfortable. Rumination times differ depending on the quality of the pasture. Shorter times are required for good quality pasture, whereas if it is fibrous and generally lower grade, the time needed increases. When calves are mixed with older cattle, they tend to spend more time grazing and less time ruminating when compared to being on their own. The weather also impacts the length of time spent eating. One study showed that cattle graze for 67 percent of the time, but windy and wet conditions reduce this to 48 percent.

Not all cattle graze; therefore, special care must be taken when storing and providing food and water. Grain should always be stored in a clean, dry environment—rodents can spread disease, so keeping them away is ideal. Milk, feed,

Below: *Feed requirements differ depending on age, breed, whether a cow is lactating, how much grazing time cattle have, and other critical factors.*

and water buckets are breeding grounds for pathogens, therefore maintaining good sterilization and cleaning routines is important. Rough surfaces such as scratches also enable bacteria to grow more easily.

Fibers such as straw and sugar beet pulp help aid digestion and rumen health. Intensive feed rations are usually carefully prepared by nutritionists to ensure appropriate levels of minerals, vitamins, and crude protein levels, in addition to general calories and energy intake; likewise the amount of sugars and starch must be monitored to avoid overweight cattle. General grazing, grass silage, maize silage, whole plant or head chop barley, red clover, rapeseed, beans, peas, molasses, cereals, citrus pulp, soybean hulls, potatoes, and straw commonly complement rations for most herds. Minerals such as calcium (Ca), phosphorus (P), magnesium (Mg), copper (Cu), selenium (Se), cobalt (Co), iodine (I), zinc (Zn), and manganese (Mn) are essential for cattle. Deficient cattle can become ill very quickly.

Whichever food is chosen, it must be stored properly in order to avoid bacterial contamination, spoilage, mold, or even loss from birds and the wind, referred to as "shrink." In general, grazing on grass is the cheapest option if available; however, the levels of minerals must be properly maintained. In the beef industry, the food type and quality will affect meat quality, especially during the finishing stages. The types of food available vary from region to region, with some crops easier to grow in hot climates while others thrive in colder, wet environments.

Left: *Feed differs greatly throughout the world depending on the local crops available in addition to formulated feeds.*

Peas

Corn

Dry beet pellets

Rapeseed

Red clover

UNUSUAL & DANGEROUS FOODS

For optimum health, cattle should be fed a carefully planned diet, and there are a number of foods that should not be given to them. Tomato plant stems can be poisonous; larkspur, nightshades, lupines, water hemlock, and poison hemlock can cause symptoms ranging from convulsions to death. Some unusual foods that have been fed to cattle include crab intestines (fishmeal), and even chocolate and candy to increase sugar levels, although these are not usually part of the diet and are not scientifically tested. Industries such as cookie factories and fruit and vegetable producers often sell off produce not fit for human consumption to cattle owners, so foods such as cookies, pumpkins, and oranges may all make their way into the diet.

Cattle also have complex rumination systems enabling them to eat different food from that consumed by people. Once crops have been harvested, cattle are frequently put out to graze the residue from corn or wheat crop fields. Cattle are often referred to as "upcyclers," as they are able to eat products such as cottonseed and dried distillers grains from brewing companies, products which would otherwise be considered waste. Years ago, chicken waste was given to cattle, but this may spread mad cow disease so is not desirable. As cattle are herbivores, they should not be fed meat.

RESTING & SLEEPING

Sleeping, drowsing, and ruminating often take up around half of the day. Cattle tend to lie down but keep their heads up or positioned back toward the flank.

The amount of time spent lying down depends on a number of factors. These include the type and comfort of their housing, their diet, the weather, and whether or not a female is pregnant.

Cattle like routine and are creatures of habit. They appear to have preferential places in which they rest, perhaps to reduce competition over differing areas, but there do not appear to be links between hierarchy and the places chosen. Rest periods are also governed by the environment. Cattle preferentially avoid noise, and will avoid their usual rest areas if disturbed. The length and type of resting periods and places chosen are also partly dependent on breed. For example, the sun affects resting preferences, with British breeds avoiding intense sunlight, whereas zebus and zebu crossbreeds remain out of the shade in order to graze, even if the sunlight is bright. That being said, most cattle will preferentially feed more at night if temperatures are high.

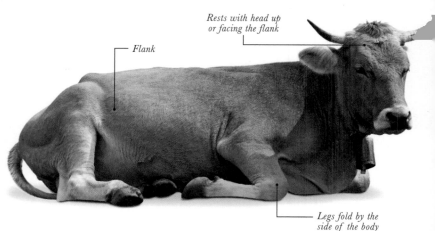

Flank

Rests with head up or facing the flank

Legs fold by the side of the body

Above: *Cattle rest, sleep, or ruminate lying down for around 14 hours daily. During these times they are more likely to stay in an upright, fairly alert position.*

Below: *Most farmers carefully control bovine diets, but as upcyclers, cattle also eat waste such as pumpkins, oranges, and distillers grains.*

Pumpkin

Barley

CAN SLEEPING COWS BE TIPPED?

Cow tipping is an urban legend whereby people sneak up on a sleeping cow and push her over, after which she cannot get up again. The tales comes from the 1970s, and even people in the Roman Empire talked about it. There are a few problems with cow tipping. First, cows can sleep quite happily on their sides and usually regain their footing after a fall unless injured. This begs the question as to why they would not be able to stand after being pushed over by a person. Farming and veterinary care can also require a cow to be put on her side, which is often called casting. This takes specialized equipment and knowledge as it is estimated that to push a cow over it would take 3,000–4,000 Newtons worth of force (the equivalent of the force of 4–14 people pushing). Cows that are standing up to sleep are highly aware of predators and so are likely to react, as might the rest of the herd. Cows can be dangerous when alarmed and generally resist against people pushing them. As well as being unkind to cows, tipping could cause injury and stress to the animal. Stress can, in turn, decrease the amount of milk a cow produces and can cause miscarriage. So, aside from being difficult, tipping a sleeping cow would be dangerous and unethical.

Breeding & Mating ✒

With artificial insemination playing such a large role in reproduction in modern farming, it could seem as if animal behavior no longer plays a role in breeding and mating; however, natural behavior patterns still have their functions and purposes. For instance, natural breeding bulls use sound as a territorial call or in competitive reproduction displays, and cows may vocalize as an estrus advertisement.

Cows usually come in heat in the evening, for 18–24 hours. At this time they may move more, eat less, urinate more frequently, show flehmen, rest their chins on other cows' backs, and lick and sniff more, in addition to mounting, showing aggression, and becoming hyperactive. Female sexual displays depend on her environment, health, breed, genes, and experience. Dairy bulls tend to mate more frequently, therefore the cows tend to be more experienced and receptive than beef breeds. It also helps if females are in a group, as estrous will be prolonged if other females are in heat. Most of the female behaviors are governed by the hormone estrogen, while the male displays are controlled by androgens such as testosterone.

A male can often detect when a female is about to ovulate—up to two days in advance—and will often spend more time around her. Sometimes teaser bulls are introduced for AI purposes, but if a bull is used for mating, he will usually try to mount the females in heat, lick at the vulva, and show the flehmen response. Standing head to head is common between males and females, as is the bull putting his head onto the cow's back.

Left: *Both bulls and cows bellow when threatened and when trying to attract a mate. Bulls also bellow when challenging other males.*

Right: *Mounting behavior is useful not only for natural insemination but also for assessing when cows are in heat, to time artificial insemination correctly.*

Timing is important for breeding, as females are only receptive to being mounted for between 1 and 18 hours per cycle, with an average time being around 4.4 hours. Generally, the most dominant males mate more frequently, but bulls can become less receptive to a known herd, therefore introducing new females increases their libido. Steers/bullocks can display sexual behaviors despite being

THREATS TO FERTILIZATION/ PREGNANCY

Fertilization and pregnancy are typically times of concern for cattle owners with respect to diseases and illnesses. For example, leptospirosis causes economic losses such as reduced milk production and poor fertility levels alongside increased abortion rates, particularly late abortions. Added to this, it is a zoonotic disease, which means people can also contract the disease; therefore, pregnant women should avoid dealing with aborting cattle. Meningitis and kidney and liver failure are also common in people contracting the bacterial infection. Venereal diseases are common for cattle at the fertilization/pregnancy period, not only when a bull is servicing a female, but also when artificial insemination methods are used. Bacteria called *Campylobacter fetus* live in bulls but can be passed to females during servicing. This venereal disease called vibriosis can cause loss of pregnancy and may go unnoticed in a herd until high numbers of nonpregnant cows or abortions are noticed.

castrated and can still achieve an erection for some time, but despite behaviors such as mounting, sexual behavior is generally inhibited. Freemartins (a sexually imperfect, usually sterile female calf twinborn with a male) are also used by producers as their increased androgens (hormones) can make them act more bullish. These behaviors are helpful to the farmer in understanding why calls are being made and to understand when a female may be at her most fertile. This can be used to time artificial insemination in females, too.

Breeding for Specifics ❧

Genetics play a very important part when it comes to factors such as milk production, meat quality, fertility, aggression, coat color, physical appearance of an animal, and overall health. With advances in genomics, many genes have been identified as playing vital roles in these factors. Each gene may have a slightly different genetic sequence, which in turn can determine certain characteristics. In cattle, the male sperm provides half of his chromosomes, including either an X or Y chromosome; the female provides half too, but can only pass on an X chromosome. Therefore, males determine whether the offspring is male (**XX**) or female (**XY**).

On the whole, genes can be dominant (the trait will be expressed), recessive (only expressed if two recessive genes are present), or can be co-expressed, and some genes are linked to other genes, for example, eye and coat color. A number of breeding tools and software programs are available to the modern breeder to help decide optimum mating. Ancestry records are often used alongside genetic and production data in order to determine potential good matches for desired traits.

Below: *Crossbreeding traditional beef and dairy breeds may result in hardy cattle able to produce more milk than beef breeds, for example.*

Pure breeding

Pure or straight breed cattle consist of a sire and dam (male and female) of the same breed parenting offspring. Although this maintains the genetic information, it can lead to inbreeding if few animals are used. There is some concern that due to intensive breeding, some herds or breeds may be very closely related, thereby lacking genetic diversity. This is particularly worrying when disease hits a herd—especially newer pathogens, as a lack of diversity and genetic variation can mean the whole breed is susceptible. Pure breeding can also reduce fertility, productivity, and longevity if poorly managed.

Crossbreeding

This type of breeding introduces new genetic traits into a herd. It can be used to develop new breeds and introduce traits such as high milk yields or high calf weights into other breeds. Heterosis occurs when the average performance of a crossbred offspring is different (usually better) to the average of the sire and dam breeds. Favorable breed combinations may come from different breed parents producing offspring who may not have better traits for one area but be more adaptable to various conditions. For instance, the offspring may be more heat tolerant; however, it may not produce as much milk as the dam (mother). Complementarity can occur when the sire and dam breeds are very different to each other but again produce progeny, which can produce a variety of traits under differing conditions. The use of a large dam with a small sire produces a large calf—another example of complementarity.

Other methods

Owners will often replace older dams with their offspring; this is classified as a continuous system. If the heifers are introduced from other herds, the system is called terminal. The sires used may be different or the same in each system, depending largely on whether AI is used or bulls on the farm.

With the data being collected and genetic data now being used, decisions are becoming more complicated. Courses throughout the world based around breeding methods, crossbreeding, use of databases, genomic data, and bull selection are becoming more popular. Nevertheless, breed and genetics are not the only things that support an animal's ability to thrive: management and environmental conditions have a large impact as well.

Pregnancy & Birth ❧

Pregnancy usually lasts 283 days, but this can range from 279 to 287 days depending on the breed, age, and condition of the mother. Cows carrying male calves will often have a longer gestation than those carrying females. In commercial units, pregnant cows will often be separated from the herd about two months before they are due to give birth. Just before she calves, the mother will try to find a dry, quiet, warm place away from other cows. Some early signs of labor are swelling of the udders, milk being released from the teats, mucus around the vulva, and restlessness. Around 6–72 hours before delivery, the mother may move away from the herd, lose appetite, feel discomfort, and show signs of contractions. Generally, a cow will take around 30–60 minutes to deliver, and heifers up to 4 hours.

Appropriate care of the mother should start during pregnancy to ensure a healthy calf. Nutrition is particularly important: poor nutrition can cause pregnancy loss, especially in cases where toxic plants are eaten or the general diet is deficient in vitamin A, selenium, or vitamin E. Natural loss occurs in around 1–2 percent of pregnancies after the 120-day stage. It is difficult to tell the loss rate before this as owners may not even be aware or

certain of the pregnancy. Producers often check for pregnancy at around day 32 post-insemination, and will recheck at 60 days, especially if a cow is showing signs of being in heat again. Levels of pregnancy loss above 5 percent are often seen as indicative of a problem in a herd. Natural loss may also be attributed to genetic abnormalities, infections, food infected with bacteria, stress, high environmental temperatures, drug-induced complications, fever, twin pregnancy, hypothyroidism, or even physical injury.

Below: *In an average breed, at day 45 the fetus measures 1 in. (2.5 cm) long, 10–12 in. (25–30 cm) at 120 days, double this by day 180, and double again by 270 days.*

Increased fat and other tissues, udder weight, the placenta, and amniotic fluid all add to maternal weight gain.

At day 250 of pregnancy there will be 20 lb (9 kg) of fetal fluid, 62 lb (28 kg) of fetus, and 9 lb (4 kg) of placenta.

By the end of pregnancy the fetus is six times heavier than the placenta.

LABOR

The birthing process begins with labor and rupture of the amniotic sac, and when all goes well, the fetus is delivered through the cervix (neck of the uterus), vagina (often called the birth canal), and the vulva. Once born, the fetus starts its life as a neonate (newborn). However, the birthing process for cows and calves can be complex, and problems crop up that need to be addressed with some urgency for successful delivery.

Dystocia is classified as difficult or abnormal birth, with around 10–15 percent of heifers and 3–5 percent of cows suffering such difficulties. An oversized calf is a problem in cattle and is more common in beef animals. This can lead to vaginal tears, hip lock (the hips get stuck), leg back and head back (where the calf is not in the correct position for birth), the calf can be delivered backward or in breech presentation, and twins may both present at the same time. Dystocia can also occur due to the calf being deceased.

Complicated labor can result in injury or death, for both the cow and calf. Careful selection of an appropriate sire can reduce the number of cases seen, as can breeding only from females that have reached an appropriate age/weight and body condition score. Frequently, interventions such as assisted delivery are required.

POSTPARTUM CARE

Following labor, the placenta is normally delivered. However, retained placenta is possible. This is more common in cases of dystocia, milk fever, and in twin births. The membranes are not usually manually removed as often the cow will expel them within 11 days. An injection of oxytocin can be given to encourage contractions, which help to expel the placenta.

Once the calf is delivered, it should be checked for congenital disorders, birth trauma, infections, and viruses. Ensuring that the mother is able to feed and that the young suckles are imperative, in order to receive the antibody-filled colostrum.

Below left: *Farmers and veterinary professionals may need to intervene during labor. Some breeds are more prone to delivery difficulties than others.*

Below: *A newborn calf will usually be licked dry by its mother. Licking prevents heat loss, stimulates breathing and blood circulation, and promotes bonding.*

Caring for Calves ⚡

Protection of the newborn starts with the mother being vaccinated, ensuring adequate intake of colostrum and milk, good ventilation, and no overcrowding in housing. Calves are most at risk in the first 28 days of life. Liquid and food loss can cause calf weight loss, which often results in poor lifelong performance. Checking the neonate for congenital disorders, breathing rates, and development is essential, and many owners will bring the herd into sheds for birthing or separate those already housed inside.

FEEDING OPTIONS

There are a number of different methods used by farmers throughout the world to feed calves. Many will use the first milk, the colostrum, straight from the cow and ensure the calf has fed within 2 hours of birth. Ideally, calves will get 3 quarts (3 liters) of colostrum a day. After the first milk, the cow will produce transition milk for 2.5 days. From 4 days old, some calves will stay with their mother and continue to feed, while others may be fed a whole milk, waste milk, or milk replacer, high in vitamins and specifically designed to support growth.

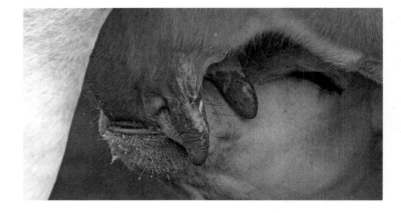

Calves will get most of their nutrients from milk in the first 2 weeks of their lives. Many will also be fed "calf starter" from the age of 4 days, which includes a range of proteins, minerals, and vitamins up until weaning. Calves need fresh clean water from birth in order for their rumen to start fermenting. Conventionally, beef cattle are weaned at around 180–210 days old, but some studies have shown that calves weaned at 120–160 days gain weight well and are still healthy. Dairy cattle are more likely to be weaned at 6–10 weeks old. By reducing milk quantities and increasing the amount of roughage and feed concentrates, the calves are encouraged to drink less milk. The weaning process usually takes 7–14 days.

Above: *Ensuring that calves feed within hours of birth and take in the vital colostrum is a key part of maintaining calf health.*

NEONATE DISORDERS

Successful feeding is fundamental in preventing many neonate disorders. Calf enteritis (intestine inflammation) levels can be reduced by making sure calves get the valuable colostrum milk; however, poor growth and death caused by enteritis still means great economic loss, around £12 million a year in the UK alone. Over the years, several vaccinations have been successfully developed that help to protect against the *Escherichia coli* bacteria that cause the illness. Calf scours (diarrhea) has numerous causes including rotavirus, coronavirus, and various bacteria and parasites. In the USA, between 4 and 25 percent of all calves born will die from scours. If the calf cannot coordinate its muscles, has a wide stance, is suffering from tremors, and not eating it could have bovine viral diarrhea virus, and these calves are usually culled. Adults can also be affected but vaccines and laboratory diagnosis tests are available.

GROWTH & HEALTH

For longer-term care, each calf should be monitored for growth according to its breed. Heart girth and wither height are measured using yardsticks or altimeter sticks with a parallel level bar. Recording difficult births and the number of deaths, maintaining daily routines, not crowding the animals, and maintaining proper handling and transportation can all help maintain calf health. Working with the youngest animals first and moving on to older animals helps protect younger cattle from disease, as can regular cleansing of housing and changing bedding materials. Keeping the calves with similar-aged animals will make caring for them easier and reduce the risk of disease.

Below: *Measuring calves helps show proper development and is also used in adults. Heart girth and wither height are commonly used to express cattle size.*

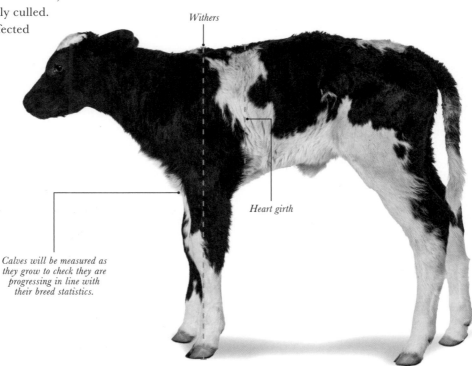

Withers

Heart girth

Calves will be measured as they grow to check they are progressing in line with their breed statistics.

Vocal Communication ❧

The general terms to describe the vocalization of cattle are mooing, lowing, bleating (generally calves), blarting, bawling (cows), bellowing (for bulls or cows when stressed), snorts, and grunts. People from differing cultures and languages express cattle noises using different words. Whether you are in Brazil or New Zealand, speaking Mandarin or German, cattle make numerous sounds, and cattle workers have an excellent comprehension of them.

CATTLE CALLS

Much research has explored whether the differing calls of cattle can be used to assess welfare and behavior. Six generalized behavior types have been suggested:

1. Lying and ruminating
2. Feeding-related behavior
3. Social interaction
4. Sexual-related behavior
5. Stress-related behavior
6. Other behaviors

Researchers have stated that vocalizations provide information on age, sex, and dominance and reproductive status of the caller, and that Asian cattle vocalize less than Western breeds.

Cattle call out to their mates and others that they are familiar with, and acoustics are specific between individual mothers and their calves (see box). Cattle generally call a lot when moving locations, and may be signaling in order to regroup. They often form strong bonds with not only their herd, but also with specific members, perhaps a close friend. Cows often vocalize when they see the farmer, probably because they suspect the arrival of food—cattle often vocalize when hungry. When a cow is lying or ruminating, she has a lower pitch than

Below: Cattle recognize herdmates' calls. They will also call out when they see people, food, if they are upset, in pain, mating, or even for no apparent reason.

Above: *Cattle often show their emotions. Whether it is tongue rolling, bar chewing, freezing, or baulking, they can display distress.*

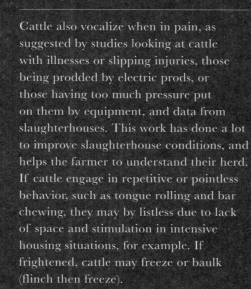

Cows often vocalize when calving, but once offspring are born, the communication between mother and calf is vital—a lack of it can indicate a problem in either cow or calf. Research has shown that individual mothers and calves have specific and exclusive calls of their own, so they know who is being communicated to. A calf will have a higher-pitched call than a cow. Bawling is frequently used to describe the sound made by a cow following the removal of her calf and vice versa. British scientists have shown that a cow may bellow when separated from her calf at a higher frequency in comparison to when the calf is nearby. The calf in turn has a distinctive sound when its mother is nearby and it wants milk in comparison to when it is separated.

normal, and in the milking parlor, you would expect the herd to be fairly quiet, indicating that they are relaxed and feel safe. Cows can also moo for no apparent reason.

FEELINGS & EMOTIONS

Cattle also vocalize when in pain, as suggested by studies looking at cattle with illnesses or slipping injuries, those being prodded by electric prods, or those having too much pressure put on them by equipment, and data from slaughterhouses. This work has done a lot to improve slaughterhouse conditions, and helps the farmer to understand their herd. If cattle engage in repetitive or pointless behavior, such as tongue rolling and bar chewing, they may by listless due to lack of space and stimulation in intensive housing situations, for example. If frightened, cattle may freeze or baulk (flinch then freeze).

In addition to fear, hunger, distress, and anger, cattle are emotional animals showing excitement and signs of depression under certain conditions. It has been said that cattle can sense death and moo at this time; many people have told of cattle getting distressed when they or their companions are unwell or dying, which cattle have shown by vocalizing. The duration, volume, and pitch of the vocalization will increase in a distressed or excited animal. The amplitude, tonality, and speed of air leaving the throat all play a role in communication.

Life in a Herd

A normal day in herd life involves standing (often while grazing), walking, lying down, drinking, ruminating, agonistic behavior including threats, fighting, retreating and conciliation, self-grooming, and social grooming. Cattle are generally kept in groups of heifers, cows, and calves, with older males kept separately until breeding. This is very dependent on the system being used, as not all herds separate males; many dairy herds separate the male calves at the weaning stage. Both males and females have hierarchical systems, with a lead bull or cow determined generally by size, weight, age, disposition, and presence or size of horns; in steers, height at the withers also has an impact. These factors all contribute toward gaining dominance but this can also be lost: for example, once a dominant female reaches old age and loses weight she can lose her position. Pregnant females tend to become less dominant due to hormonal changes. In general, males will dominate over females; younger males can gain dominance as they put on weight. Even juveniles have their own hierarchical systems among peers. Once dominance has been established, a slight movement is enough to cause submission in another animal.

A very large herd often splits, enabling differing animals to lead the hierarchy for each of the smaller groups. Normally, food is shared, but studies have shown that when access is only available to one or a few animals, the more dominant animal will control the stall. When living in mixed-breed situations, the heavier cattle habitually gain dominance in dairy breeds, whereas in beef animals, the lighter breeds may be dominant. Although hierarchies may change depending on which animals join or leave the herd, once dominance is achieved, the herd tend to be less aggressive overall.

Milk production also influences how the herd interact with one another and their surroundings. Cows producing higher milk yields tend to enter milking parlors earlier than those with lower yields. Milk production can decrease in cows that are newly introduced to a herd while they establish their position within the hierarchy.

Grazing is a fairly structured affair in which younger cattle remain in the middle of the herd, whereas sick, weak, and aging cattle are more vulnerable to predators as they fall behind the herd or remain on the outer edges.

Right: *Whether following the leader, Earth's magnetic lines, or simply facing threats and avoiding in-herd confrontation, cattle tend to face the same way while grazing.*

FOLLOW THE LEADER

The leader of the herd is not necessarily the dominant animal, neither does it truly lead the direction of the herd. Some have suggested that dominant cattle stay in the center of the herd, while other studies indicate that higher ranking cattle lead and lower ranking ones follow. When deciding the direction of travel, the lead animal may initially decide upon the direction, and if the herd changes direction, the lead animal will simply move to the front and take the lead again.

A MAGNETIC ATTRACTION?

A study using Google Earth satellite images showed that cattle are more likely to face in the north–south direction of Earth's magnetic lines. The wind direction in general and time of day did not affect this tendency, but a very strong wind or sunlight on a cold day could make cattle change their direction. This study could not see the animals when they were indoors, but it certainly suggests that cattle have a sense of magnetic direction. Other studies have refuted this work, suggesting that magnetoreception is not a bovine sense at all.

Behavior & Temperament ⌘

Cattle are generally contented feeding, chewing the cud, ruminating, and being with their herd. They settle into a hierarchical system and like routine and habit. In general, they are fairly docile animals and are unlikely to fight with one another or people. They are frequently moved around ranches by people and dogs, and show little resistance to this.

Some cattle behaviors are instinctive—they are generally nervous of new environments or things in their vicinity, and if a cow becomes afraid she can take around 20 minutes to calm down, potentially instilling fear into the rest of the herd. As prey animals, this alertness is a primary method of avoiding predators and is, therefore, a natural and understandable response.

Below: Cattle are naturally curious, but as prey animals they can also be cautious and nervous, especially when faced with new environments, people, and animals.

Cattle are often associated with aggression, and in a herd, cows often show this in a sequence of approach, threaten, physical contact, and fighting. Once cattle have established dominance, the aggression tends to subside, and the loser becomes more submissive, avoiding the dominant leader and assuming a passive posture. It has also been suggested that reactive cattle are more likely to have high hair whorls on the face.

Males may display their size, and therefore dominance, by turning their body perpendicular to another animal, thereby showing their full height and length, while others may turn toward another male and bunt or strike using their head. Bunting is not always aggressive. It is often seen in young calves as they bunt their mothers' udders in an effort to stimulate milk production. Both young and older cattle can also use the action to play. Bunting is useful in self-defense; for example, by a less dominant member when attacked by an alpha in the herd. In addition, a more dominant animal is more likely to bunt another in the rump; when the head is between the legs of the more submissive animal this may be referred to as clinching.

Above left: *Some studies have suggested that high hair whorls on the head indicate temperamental behaviors and aggression.*

Above: *Bovines such as bison, buffalo, and domesticated cattle breeds often use bunting and fighting to show their dominance.*

BULLS & PEOPLE

The now renowned bull runs stemmed from the tradition of simply moving herds from one place to another, for breeding or selling at market. Over time, young people would run in front of the cattle or jump on them as a sign of bravery, and this became a celebrated practice. Many human injuries and even deaths have arisen as a result, but few are from being gored (pierced with the horns)—most are running or trampling injuries, especially if the streets become blocked. Cattle have died and been injured, too, often from stress, and many people now campaign against such events on an ethical basis (see page 156).

Bullfighting is a cultural sport in which the cattle are immobilized, subdued, or killed. Early examples appear in Greek mythology of fights and the killing of sacred bulls, and the Romans frequently used these powerful creatures instead of gladiators at their games. During medieval times, the Spanish considered it a noble sport. Modern bullfighting employs matadors and their assistants (picadors and banderillos), sometimes on horseback. Traditionally, the bulls were lanced and pierced, and finally killed with a sword. There are a variety of additional rituals which are used to weaken, tire, and agitate the bull. In some places, bullfighting is performed without physical harm to the bull. The Spanish, Portuguese, and French have a history of bullfighting, but variations of bull-centered "sport" happen throughout the world: Jallikattu in India, bloodless fighting in Tanzania, and events where bulls wrestle each other.

Above left: *For the Pamplona running of the bulls festival in Spain, certain rules are now observed, such as not inciting the bulls and the humans getting a head start.*

Above: *Traditional matadors have brightly colored capes, or* muleta. *Bulls do not see the color red as they are dichromatic, therefore they react to movement.*

There is no doubt that bulls are generally large animals that can run at over 25 miles (40 km) per hour. They can also be more aggressive when not castrated. Bulls are likely to react in response to being provoked; their general environment and the way they have been treated will dictate how aggressive or calm they are. Bulls usually only make up around 2 percent of the cattle population worldwide but cause around half of the injuries to workers/owners. Bulls frequently fight in the wild in order to become the head of the herd. Unusually, hand-reared bulls are more dangerous than herd-reared bulls, as they may challenge humans rather than other bulls in their quest to become the boss. A bull will often show signs that he is about to attack, such as arching his back, stomping his hooves on the ground, and lowering and shaking his head, but he often stops the aggression if the person or animal backs away at this stage.

Fear levels, hormones, upbringing, and environment all affect mood. Understanding behaviors is essential for cattle handlers, as the animals may bunt, kick, or crush people, too. Being able to recognize signs of aggression, threats, displays, and retreats helps workers stay safe—especially when there are sizeable horns present. Generally, within the farming community aggression in cattle is avoided; where possible, by changing breeding patterns, the environment, and by training the animals, but where necessary, the animal may need to be culled.

DO CATTLE CHASE PEOPLE?

Cattle are generally inquisitive, but can be fearful of people, animals, loud noises, and machinery. Upon entering a field of cattle, they may not react at all, but if they do, it could be out of interest, fear, or simply in the belief that you have food. How they behave often depends on how much contact they have with people in general. Other animals such as dogs may also cause the cattle to worry. Running toward a herd or causing widespread panic by shouting and behaving erratically is not advisable and could cause the herd to charge. Like most mammals, a cow will instinctively try to protect her young, therefore approaching calves may be dangerous. Agitated cows may shake their heads, snort, and move their tails; a bull will snort and paw the ground. Calmly moving away from the animals is usually a good idea if they are displaying signs of distress.

Intelligence, Learning & Memory ❧

Cattle can become accustomed to routine, which shows intelligence, learning, and memory, as entering the dairy parlor and finding food in familiar places are learned behaviors.

Spatial cognition can play a part in speed of grazing; for example, cattle that change paddocks regularly are more likely to eat faster than those who stay in them for longer. Those rotating more frequently also withstand lower food levels, as they understand that fresh opportunities will be available in the next rotation. Changing paddocks may also initially increase the amount of time exploring the pastures rather than concentrated grazing time.

In short-term studies, both heifers and steers recalled directions to food within simple mazes. Their memories were best within 4 hours, declining significantly after 8 hours. In more complex mazes

Below: Cattle grids are a well used method of preventing animals moving out of territories, they will even stop at painted lines on a road.

over more time, cattle have recalled food locations for up to 6 weeks. Steers have recalled the placement of food buckets for at least 48 hours, but when faced with familiar stimuli, cattle can associate food buckets with food for up to a year. In repeated tasks, cows show long-term memory abilities, recalling the locations of food for at least 6 weeks in familiar territories. In addition, they show spatial cognition by systematically searching for food rather than randomly strolling around. In a task using alarm bells, the cattle reacted after just seven trials of the sound and had appeared to understand the meaning of the noise.

Cattle will also remember obstacles, such as cattle grids, and will avoid them. When fake grids are painted onto roads, they avoid these too, but if put under pressure to cross, or presented with a reward, such as food, some animals will cross over then others will follow. Cattle not previously exposed to cattle grids also avoid the fake ones, therefore it is likely the change in color and pattern worries them. Before all ranches paint gridlines, it is worth remembering local conditions—a muddy covering will render the fake grids useless. Cattle have even learned to switch housing lights on and off by putting their muzzle into a light beam, and feed themselves sodium solutions when they are salt-deficient.

MEMORY & RECOGNITION

Object discrimination is a principal feature of memory and learning. Cattle can differentiate between differing flowers and grasses, even when they are just pictures rather than the real thing. Shape, size, brightness, and color all appear to be discerned. It has been suggested that cattle have long memories when identifying individuals within a herd of between 50 and 70 members, which is, perhaps, why larger herds break into smaller groups, to enable recognition. Studies show they recall fellow cattle for at least 12 days, likely a lot longer in reality. They even remember specific dogs, elephants, and pigs, while simultaneously understanding they are a different species to themselves.

Below: *Identifying fellow herd members, other animals, and familiar people shows both recognition and memory skills.*

EMOTIONS & MEMORY

Emotions may play a role in recognition and memory. Cattle can recall or react to situations which were previously positive or negative, such as food, treats, or pain. They remember handling methods employed by people and recognize the people interacting with them, even if the people are wearing the same clothes as one another. Emotional contagion is thought to be a basic form of empathy, and certainly cattle can react to other individuals' emotions, especially fear and stress. They often become stressed, eating less and producing more of the stress-related hormone cortisol, when others do. Although they may detect one another's emotions, cattle can also be calmed by being near others, especially those in their close social circle, but also by the presence of a mirror, their own reflection mimicking being with another herd member. Play is often associated with happiness in cattle and, in turn, they respond positively to it. They can learn play and other tasks from one another, which shows good mental capacity. Indeed, when cattle master a problem, they appear to show signs of excitement, such as increased brainwaves and heart rates, indicating self-awareness in addition to intelligence. Finally, cattle have distinct personalities. Whether this is genetic, environmental, learned behavior, or a combination of all three, each animal has its own traits and quirks.

Left: *Cattle will often gallop, twist, and turn, and play with soil, straw, and other objects. They will even play fetch with a ball.*

DO COWS LIE DOWN WHEN RAIN IS FORECAST?

British people in particular love to talk about the rain, and apparently 61 percent of the population believe that a cow lying down means a higher chance of rain. There are theories that purport that cattle lie down due to atmospheric pressure changes or increased moisture in the air as a storm approaches. There is also a theory that increased water in the air goes in through their skin, forcing the cow to lie down. One study from the University of Arizona and North Missouri suggested that air temperature changes may make a difference, as cows are more likely to look for shade and stay standing when hot in order to cool down. So far, there is no scientific evidence to prove that our bovine friends can predict storms or have memories of when storms occur. They are probably just lying down to chew the cud, having a rest, or suffering from heat stress, pain, or illness, or they may just be following the herd leader. It's worth noting that up to 30 percent more blood circulates through the udder when a cow is lying down. So, don't get out the umbrella every time you see a herd lying down—there are a host of other reasons more likely to be behind their reclined positions.

Cattle Disorders & Injuries ✦

The skin & feet

The skin is an important self-defense mechanism, helping provide a physical barrier from the outside world and helping to prevent pathogens from entering the body. However, skin is also prone to disorders and injuries. Foot rot is a common infection originating from lesions in the interdigital skin. Exposure to wet conditions, piercing, or thinning of the skin can result in bacterial invasion and accounts for around 15 percent of claw diseases each year. Other causes include digital and interdigital dermatitis, especially in dairy cattle; the former can be difficult to control as it is so infectious. Sole ulcers can form when the skin underneath the horn is damaged and dies, and laminitis forms when the layer called the corium becomes inflamed. A hoof disorder called "white line defects" often occur due to foreign objects such as stones damaging the more delicate tissue, and chronic irritation of the skin can also lead

Above: *Feet frequently get infections, and foot rot is a common disorder. Keeping the environment dry and clean, and disinfecting feet, are key preventative techniques.*

Internal organs

Bloat is more common in beef cattle grazing, especially those eating legumes. The build-up of gas from the fermentation process causes pressure on internal organs and can result in death. Displacement of the abomasum is also common in cattle due to a build-up of gases. Abomasal ulceration can be common, and can result in gastric bleeding and even perforation of the tissue and cause death. Ingestion of hard, foreign objects such as nails and glass can also cause extreme damage and perforation to the intestine and other organs. These types of injuries and disorders lie alongside those linked to a lack of nutrients and minerals, and general feed-related issues including bacterial infections. Bovines can also get cancer, such as enzootic bovine leukosis in adults and sporadic bovine leukosis in the young.

Left: The skin around and near the hoof is particularly vulnerable to infections and disease, largely due to close and constant contact with the ground.

to growth of tissue in the interdigital space called tyloma. The use of nonsteroidal anti-inflammatory drugs, foot baths and cleaning, use of blocks/shoes, removal of dead tissue, antibiotic sprays and injections, clean and dry living quarters, and regular foot checks and mobility scoring are essential preventive, identification, and curative treatments. Hoof trimming is also essential in reducing foot disorders. While there are ongoing studies to determine the optimum trimming protocols, the reduced welfare levels and milk yield losses from cattle with foot problems are well documented. For example, studies from Finland, the UK, and the USA indicate milk losses of 3–6 lb/day (1.5–2.8 kg/day) per cow can occur.

Right: This cow has a rumen fistula hole to help treat indigestion problems. Also referred to as a cannula, the first documented use was in 1928.

Skeletal system

Bone health is an issue for most mammals, but in cattle undergoing rapid growth and often high lactation levels, frail or broken bones can be a common problem. Diet plays a key role, and genetics and production are important in other disorders. For example, pregnant cattle fed on silage over winter can give birth to deformed calves due primarily to manganese, which can cause bone deformities and enlarged joints. Marble bone disease, also known as osteopetrosis, is passed on via genetic lines, especially in breeds such as the Hereford, Angus, Simmental, and Holstein. Both parents need to pass on a mutated gene so that offspring have two mutated versions which prevent bones cells called osteoclasts from breaking down old bone, and also stops them from helping to develop and remodel new bone. Deoxyribonucleic acid (DNA) tests developed from the discovery of the mutated gene in 2010 help identify adults carrying a defective copy of the gene, thereby preventing breeding between males and females who both carry a mutated gene. With marble disease-affected calves suffering from brittle bones, being stillborn, or dying within a day of birth, eradication of the mutated gene is desired. Other forms of chondrodysplasia (bone and cartilage disorders) in cattle include dwarfism, "bulldog calves," with severe malformations of the bone, and osteogenesis imperfecta. Arthritis and fractures can also occur due to everyday wear and tear and accidents.

Some muscular problems such as muscular hyperplasia, for example, in double-muscled animals, and crooked tail syndrome often occur as a byproduct of increasing general muscle mass in breeds. Many musculature problems lead to other side effects, such as heart and nervous system failures and low reproduction rates. Nerve damage may occur as a result of many situations, but a cow going down (refusing to stand) can cause nerve and muscle damage. Trauma, mastitis, milk fever, lack of calcium, and a number of illnesses, injuries, and dietary issues can cause this problem. Difficulties with the nervous system can even be caused by medical treatment such as injections and surgery; therefore, the latest research highlights where injections are best placed, to avoid such injuries.

Below: *"Speckles" is a 2-month-old dwarf calf. While some dwarf cattle live a good life, others may encounter severe bone problems.*

General hygiene & habits

Social standing, general hygiene, and habits also contribute toward injury levels and health. Social grooming, termed "allogrooming," is most frequent when food has been delivered to a feed bunk or feeders, which might be a tension-reduction technique related to competition for food. Another common allogrooming time is between midnight and 2 a.m. If there is a lack of food, this behavior declines, to the detriment of lower-status cattle in the herd, especially primiparous cows (only given birth to one calf or pregnant with their first calf). This can make younger females more at risk of poor coat hygiene. It has also been suggested that allogrooming reflects friendship, as cattle were more likely to groom their partners at the feeder.

To maintain good health, cattle must have the ability to walk and lie down effectively, which relies heavily on space allocation and the type of housing or land the cattle are on. Agnostic behavior can affect the general welfare of the herd, so it's important that cattle's basic needs are met (see page 85) to limit this.

Cattle tend to show pain less than people do, and pain is generally lowered when the animal is alongside its herd mates. It may be expressed through an abnormal stance and gait, with tucked abdomen and tail, hunched back, unusual walking, and standing still for longer than normal. Cattle may also change their resting positions, vocalize more, kick more frequently, and swish their tails. They may eat and ruminate less, and even grind their teeth.

Below: *From the moment they are born, calves are washed by their mothers. Social grooming is important in calves and adults for bonding and hygiene.*

Cattle Disease & Illness ✧

Like most other mammals, cattle can suffer from a wide range of diseases and illnesses, which differ in levels between farms, regions, and countries. Covering the thousands of illnesses and diseases is impossible, but understanding important self-defense mechanisms, common diseases, and general traits is essential.

Below: *This French herd are receiving their blue tongue vaccination. Prevention methods also include insecticides to kill the vector, restriction of animal movement, and slaughtering infected animals.*

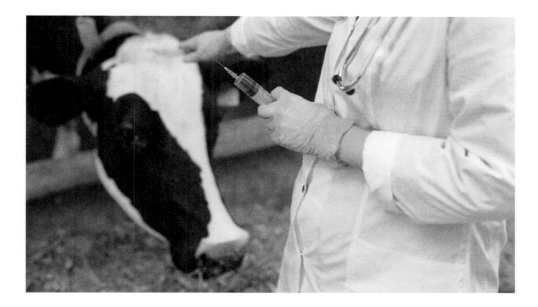

Above: *Many different types of cattle vaccinations have been developed against a range of diseases.*

DEFENSE & VACCINES

Cattle have a vast range of defensive mechanisms throughout their bodies in order to evade infection and disease. White blood cells and antibodies work together to detect foreign bodies such as bacteria and destroy them.

Vaccines and other drugs are available for many diseases, to help the body prepare for viral, bacterial, and parasitic invasion, so that it recognizes these pathogens early and mounts a suitable response. Types of vaccines include:

Live, attenuated vaccines—introduce a mild but symptomless form of the bacteria or virus, and inactivated vaccines, which contain destructive agents such as bacteria or viruses that are killed before being introduced to the body.

Toxoid vaccines—use dead toxins from pathogens.

Subunit/conjugate vaccines—insert specific proteins or carbohydrates into the cattle, enabling them to react if bacteria or viruses with those traits enter the body. Some bacteria coat themselves with a sugar coating in order to camouflage themselves; conjugate vaccines teach the body to recognize this coating, enabling them to destroy the invading bacteria.

DNA and recombinant vector vaccines—still new and in experimental stages, but these use DNA from the harmful pathogen, enabling the body to react against the invasion.

As pathogens can evolve and adapt, vaccinations must be developed and updated over time to ensure that all pathogens, or as many different types as possible, can be targeted. More than 40 vaccines are licensed for cattle throughout the world, and for some diseases such as Bovine Viral Diarrhea (BVD) several different types are available. This virus can cause economic losses of $2.40–687.80 per cow around the world, which differs between different regions and farm types.

COMMON DISEASES

Pneumonia
Affecting all cattle—young and old—at considerable cost, pneumonia causes weight loss, demands extra labor and treatment, and can result in death. While building design and ventilation are essential in reducing the number of cases, immunization can also help to prevent illness.

Clostridial disease
Bacteria live in a variety of places, from soil, grass, water, and food, to feces and carcasses. Clostridial disease is no exception, and can cause a number of illnesses, from tetanus to botulism. Calves are especially vulnerable, as branding, dehorning, having injections, castration, poor housing, and even movement of animals offer extra opportunities for pathogen invasion.

Pink eye
All diseases vary by region, but some are very particular for environmental types or vectors that are more dominant in some parts of the world. Examples of this include infectious bovine keratoconjunctivitis (IBK, often called pink eye), a contagious disease prevalent when the weather is very dry, such as in Australia.

Bovine ephemeral fever
Also known as the "three-day sickness"), bovine ephemeral fever (BEF) is caused by a virus passed on via mosquitoes. In parts of west Africa this disease is endemic; therefore, vaccinations are also suggested for neighboring areas.

Above: *This Hereford steer has infectious bovine keratoconjunctivitis. Also called pink eye, New Forest eye, or blight, this bacterial infection is spread by contact or flies.*

Tuberculosis
TB in cattle is caused by the bacterium *Mycobacterium bovis*. Not only is it zoonotic and can be passed on to humans in raw milk, but controversies surround the disease in relation to wildlife. Badgers are known to carry *M. bovis*, but whether they spread it directly, indirectly, or at all to cattle, and to what level, is a matter of debate. Despite vaccinations being available, it is difficult to vaccinate badgers, and in some areas badger culls have been trialed. Scientific investigations are still underway to understand routes of infection. Infected animals can pass on TB via breath, milk, discharging lesions, saliva, or urine, and feces, infected food, and water are also a problem. Farms at risk of TB often test for the disease and cull affected individuals to protect workers, the human food chain, the herd, and nearby herds. Cases of bovine TB in humans are much higher in the developing world, primarily thought to be due to the increased use of unpasteurized dairy products in those regions.

WORLDWIDE REPORTABLE DISEASES

Some diseases pose such a threat to worldwide cattle that they are reportable to The World Animal Health Organization.

Bluetongue

Caused by the virus Reoviridae, and spread via midges, historically bluetongue was confined to tropical and subtropical areas but it is spreading. It is especially difficult to prevent infection as the midges continue to harbor the virus, and the virus itself evolves at a rapid rate. Typically much higher in sheep, infection levels and fatalities in cattle are at a lower level, both being under 5 percent. Intervention programs include killing the midges with insecticides, using mosquito nets, vaccinations to protect animals, and not transporting affected animals.

Rift Valley fever

Discovered in Kenya in sheep and cattle, Rift Valley fever is now on the worldwide reportable diseases list. Like bluetongue, it is carried via insects, specifically

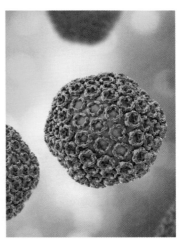

mosquitos, causing death in young and abortion in 100 percent of pregnant cows. As a zoonotic it can spread to other species, including people, where although the symptoms are often mild, they can be severe. Human infection from cattle is airborne or via infected meat, raw milk, and blood. With no cure for people or cattle, the fever can be dangerous.

Anthrax

Caused by the bacteria *Bacillus anthracis*, anthrax can infect most mammals, although ruminants and people are more susceptible and can die within hours following infection. Prevention by avoiding food and air containing spores is essential, as are vaccinations in affected areas; on rare occasions antibiotics can help treat affected individuals, but this rarely works due to the quick death caused. Russian scientists estimate that around 1.5 million anthrax-infected reindeer carcasses reside in the Arctic permafrost, and spores are known to live at least 105 years even in those extreme conditions.

Above left: *The* Aedes aegypti *mosquito spreads Rift Valley fever in cattle. It also transmits Zika, dengue, chikungunya, and yellow fever to humans.*

Above: *A magnified view of the virus that causes Rift Valley fever, infecting animals and people.*

Below left: *The bacteria* Bacillus anthracis *is grown in a laboratory, where research is undertaken into understanding and preventing the deadly anthrax disease.*

Bovine spongiform encephalopathy

BSE drew considerable media coverage in 1986, when the fatal brain disease first emerged. The prion (a type of protein that causes proteins in the brain to fold abnormally and not function) which causes it is stable and resistant to high temperatures, disinfectants, and UV radiation. Found in feed containing bovine brain and spinal cords, it killed cattle and quickly became known as "mad cow disease." It is now prohibited to add bovine carcasses to feed, and offal for human consumption has been outlawed in many countries due to the strong links between BSE and Creutzfeldt-Jakob disease (CJD), observed in people. Of the four types of human CJD, one called "variant CJD" has been associated with people eating offal, with three cases via blood transfusion transmission. Although cases are rare, with around 230 patients identified with this variant type worldwide, there is no cure for CJD, and the prion causing it can lay dormant for over a decade; yet with the onset of symptoms, death is observed normally within 14 months. This disease highlights the importance of good practice in farming and food industries.

Foot and mouth

A highly contagious, airborne, viral disease, foot and mouth is easily spread between cattle when sharing food, or moved by vehicles and people traveling around on farms. It is heavily controlled across borders and differing countries. Infected countries are not usually allowed to transport animals—the most efficient method of further outbreak is to cull all animals and disinfect entire farms. Regions work hard to remain disease-free without use of a vaccination but can also achieve disease-free status with vaccination use. Despite 77 percent of the world's livestock carrying the disease, most of Europe, North America, Australia, and New Zealand have retained their disease-free status. The USA saw a vast outbreak in 1914, with 170,000 animals contracting the disease, costing over $4.5 million, an extremely large sum for that era. The UK saw a devastating outbreak in 2001, with around 2,000 confirmed cases leading to slaughter of more than 6.5–10 million sheep, cattle, and pigs across 70 percent of farms.

Below: *This cow is showing the symptoms of foot and mouth disease. Highly contagious, it must be controlled. Extreme losses to life and income result from such infections.*

Costing over £8 billion, the cull caused tourism rates to reduce by 10 percent, increased cases of suicide and mental health problems in farm owners, but also ensured foot and mouth eradication in Britain. The Chinese reported an outbreak in 2005, and their agriculture ministry is still reporting cases in differing areas. In 2010 and 2011, Japan, South Korea, and North Korea reported outbreaks. By 2012, a global strategy was implemented to help control and understand foot and mouth. Since the discovery of the virus that causes the disease, in 1897, the millions of deaths and slaughters, loss in productivity, cost of treatment, and funding of eradication programs are estimated to have cost up to $23.5 billion per annum worldwide. Foot and mouth cannot be contracted by people; an unrelated, usually mild disease called hand, foot, and mouth is common in children and the two are frequently confused.

Infectious bovine rhinotracheitis

Caused by Bovine Herpesvirus-1, which can result in respiratory and reproductive system infections, infectious bovine rhinotracheitis can lead to fatal pneumonia, but economic losses are more usual, in the form of reduced milk production and fertility coupled with increased numbers of abortions.

Diseases affecting cattle, and in some cases people, highlight the importance of cattle care, legislation, transport laws, using appropriate handling precautions for animals and their products, treatment, prevention techniques, and veterinary public health. Veterinary professionals often concentrate on animal care and support for owners and workers, but they also all work together to ensure food and product safety by checking food and milk, slaughterhouse conditions, equipment, adequate disposal of undesirable produce, and ensuring safety of food and water given to animals.

Above: *Carcasses were disposed of on farm pyres during large foot and mouth outbreaks such as those in 2001.*

Maintaining Good Health & Condition ✀

Housing conditions, appropriate breeding, and good food and water supplies all help maintain health and body condition. In addition, veterinary and medical treatments ranging from hoof trimming through to the use of preventive treatments and pharmaceutical drugs such as anti-inflammatories, anthelmintics, antibiotics, and vaccinations help to maintain herds. Some producers may also use drugs such as steroids to promote cattle growth, but countries have differing rules on whether these can be used in products for animal/human consumption. Following the use of some drugs, animals may need to be quarantined and need extra care and time spent on their welfare checks. In many cases, especially with anthelmintics, the meat and dairy products cannot be consumed by people for periods of time ranging from none to a few hours or days

Below: *Each cattle owner must carefully design their herd's healthcare regime for optimum health, as displayed by this Devon cow. Each breed's needs will differ from other breeds and also depends on the specific environment.*

(more likely with milk), up to many months (more often in the meat industry) following treatment.

A number of preventive and curative pharmaceuticals, including anthelmintic (also called antihelminthics or deworming) treatments, are available to farmers. These generally help against worms such as roundworm, tapeworm, lungworm, and fluke. Anthelmintic resistance is increasing in many animals, such as sheep and horses, and is a concern in cattle, although it's less common than in other species, especially when appropriate strategies and protocols are followed by veterinary professionals and owners. Treatments range from injections, pour-on solutions, and bolus preparations, through to solutions in feed or given via oral drenching.

THE VACCINATION DEBATE

While some people may not believe in vaccinations, they reduce the use of antimicrobial drugs, increase herd productivity in areas such as milk, meat, and working ability, reduce illness, and increase fertility. Some common problems on farms and ranches are that secondary doses of vaccinations that are sometimes required may not be administered, are given at the wrong age, or are kept in conditions unsuitable for that type of vaccine. In some economic areas, the cost of the treatments or preventive therapies such as vaccinations may outweigh the cost of loss production, making it unviable to treat or prevent. Vaccination rates fluctuate around the world, but for some diseases, such as pneumonia, vaccination rates have increased dramatically. Around 80 percent of large

livestock producers in Europe, the USA, South Africa, Australasia, and East Asia vaccinate their animals, due to the severe economic losses observed when nonvaccinated animals contract the respiratory disease. In backyard and smallholder farms, especially in sub-Saharan Africa, Asia, and South America, the number of vaccinations can be extremely low and difficult to quantify.

Some herds or countries do not have any incidences of particular diseases, and so decide not to vaccinate. This enables them to sell cattle to differing areas which are also disease free. Therefore, overall herd health is dependent on the market prices for dairy products, meat, and other goods, alongside affordable pharmaceuticals and treatment methods, the costs of food, water, housing, and other materials required, and the cost of labor. This must all be offset against the cost of not treating animals, lost productivity, treatment, care of sick animals, and even culling and disposal costs of nonvaccinated cattle that die.

Above: *Environments, housing, diseases, and water supplies differ around the world. Ringworm is one of the most common skin diseases in cattle and is caused by a fungus.*

THE ANTIBIOTIC AGENDA

Antibiotics and steroids are also used in many farming systems. With antibiotic resistance becoming a real concern in both animal and human medicine, appropriate use of these drugs is becoming a worldwide agenda. In a similar manner, the use of steroids/ hormones is being reduced in many systems, not due to resistance but more in relation to pollution and human health. So although steroids/hormones may help cattle to increase weight or help increase milk yields or reproductive rates, differing countries have diverse regulations on which substances can be used in production animals. It should also be noted that some steroids are used to treat illnesses, rather than to increase efficiency and profits alone.

WORKER WELFARE

Providing good health and conditions for cattle essentially comes down to two overriding factors: time and money. There are many economic factors, often outside of the control of individual farmers, which impact on the ability to provide adequate feed, space, housing, medicine, and veterinary care. While farmers provide the best care possible, it should be noted that good health and conditions for farmers and their workers should also be considered, especially as these directly impact cattle care. Farm work can be extremely dangerous, and seasonal workers who frequently have less training and experience in cattle health and behavior, as well as general health and safety in agricultural environments, are particularly at risk.

Above: *Antibiotics have advanced the world of medicine and healthcare, but antibiotic resistance has become a threat to both people and animals.*

Suicide, depression, and other mental health concerns are often highlighted in the cattle-farming community as they frequently have high rates across the world, and disease outbreaks are a real source of financial deprivation, grief, isolation, and concern for those involved. These reasons are added to general debt, corruption, poor decision-making around government policies, access to dangerous chemicals, and even high altitude as links to depression which may affect farming communities around the world. In India, 11.2 percent of all suicides are farmers; it's 1 percent in Britain; France sees a suicide rate 20 percent higher for farmers and 30 percent higher for dairy farmers than the general population; and while the data is somewhat incomplete in the USA, "Farmers, Ranchers, and Other Agricultural Managers" had double the suicide rate of the rest of the population. Therefore, farming ranks as one of the most dangerous industries to work in and has one of the highest suicide rates.

These factors may also impact on a farm's ability to provide enough healthy workers throughout the year. Rural areas may not be able to provide enough foot trimmers, for example, or veterinary professionals to help farmers with their herds. Injuries, deaths, suicides, and all of the contributing factors such as finances, poor equipment, training standards, and availability of workers and healthcare providers, will have an effect on animal welfare.

Below: Farmers and workers need to be trained to use modern machinery safely—not only for the sake of the cows but also for their own welfare.

CHAPTER 4

Cattle & People

Cows for Profit:
The Economy of Cattle ❧

Cattle play pivotal roles in the economies of regions worldwide. Their uses range from helping a family to survive on a small rural farm by draft work, to providing meat or dairy products for a family. On a larger scale, their roles could be being part of a large dairy or beef farm or a stud farm, with larger numbers of cattle. There are also many related industries which depend on cattle, such as veterinary, healthcare, and artificial insemination industries, pharmaceutical companies, food processing plants, leather and hide factories, clothing industries, and many more. Many livelihoods depend on cattle.

Below: *From leather shoes to pharmaceutical gelatin capsules, products derived from cattle are an important part of most economies.*

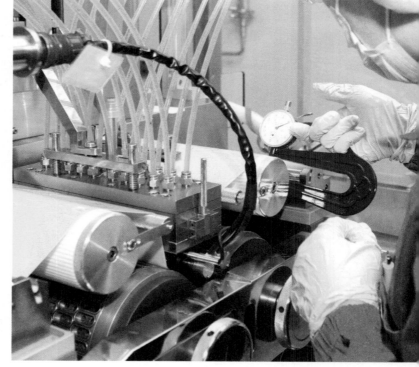

BEEF

Commercial beef production forecasts for 2019 in the USA alone indicated that 27.61 billion lb (12.52 billion kg), would be produced, with steer prices ranging between $115 and $122 per 100 lb (45 kg) (termed a hundredweight or carcass weight, cwt, in the industry). In context, the total red and poultry meat production in the USA was 102.4 billion lb (46.4 billion kg) in 2018. Meat production is increasing, with the average US citizen consuming 57.2 lb (25.9 kg) of beef products per annum in 2018. Beef production and consumption in many European Union countries has declined a little over the last few years, with a 2019 per person consumption rate of 23.8 lb (10.8 kg) per annum. Consumers can expect to pay different costs for beef depending on the cut and quality.

In 2019, average prices of beef round steak were topped by Switzerland at $50.05, followed by Norway and Iceland at $26.83 and $24.62, with Colombia and Pakistan near the bottom of the list with $4.27 and $3.02 respectively.

MILK & DAIRY

The Animal Health and Dairy Board, based in the UK, showed that around 210 gallons (794 million liters) of milk were produced worldwide from cattle per day as of January 2019. Twenty-eight European Union countries, Argentina, Australia, New Zealand, and the USA produced 65 percent of this amount between them. Other species such as goats and other bovines also produce commercially available milk, including 108.7 million buffaloes and a few million other bovines.

Above: Many billions of tons of beef are produced every year for human consumption, such as burgers and steaks, and to feed other animals such as cats and dogs.

The International Farm Comparison Network (IFCN) compares farm level data in 53 countries and economic data from 115 countries. This represents 98 percent of the global milk production. Its work helps highlight the cost of dairy production, how much dairy products can sell for, and how prices of related products, such as cattle feed, can influence the industry. IFCN data shows that world average production prices of milk reached $40.20 per 220 lb (100 kg) in 2018 and fluctuated around $33–42.7 in 2019. Variation in production prices around the world went from $20–105. In reality, this meant producers could get a profit of $2.8 per 220 lb (100 kg) of milk on average. Research highlights that the dairy industry has 7 billion costumers, and demand is increasing. Long-term

trends show that between 1998 and 2016, milk production increased by 2.3 percent.

The dairy industry produces milk, cheese, whey, and other dairy products that are all key factors when considering the economics of dairy cattle. In 2019, the British Cheese Board had over 700 types of British cheeses alone on its list. The first cheese factory dates back to 1815, in Switzerland, but it was the USA that saw the first large-scale cheese production lines. In 2018, Forbes and

MUSICAL MILKING

Believe it or not, there may be some truth in the belief that playing music to cattle increases milk production. Cows produced up to 3 percent more milk when played REM, Simon & Garfunkel, and slow, rhythmic music in general, including classical music. They also liked a bit of Shakespeare, but showed no reaction to fast-paced music, so those club anthems are out. This is not entirely surprising. In order to produce milk, cows rely on a hormone called oxytocin, which is more likely to be produced when the cow is not stressed. There have even been competitions for people to create the best music for cows and suggested playlists for the discerning music-loving farmer and cow. There is even photographic evidence of The Ingenues, an all-girl band and vaudeville act, playing to cows at the University of Wisconsin back in 1930. Many a farmer will tell you that their herd loves the radio, and, of course, increased milk production can result in better economic return for a farm.

many other media outlets highlighted the most expensive cheeses in the world. British White Stilton Gold came in at $400–420 per lb, and Pule Balkan donkey milk cheese came in at $600 per lb. A cheese from China came a close fourth, which is impressive for a nation that eats only 0.074 lb (34 g) of cheese per person per year, and an Italian cheese came in at fifth most expensive. With the average American eating 38 lb (17 kg) of cheese a year, and Denmark topping the charts with 62 lb (28 kg) per person per annum, life could get expensive if all cheese cost that much. It is also a surprise that French cheeses did not make the list, as the French have a long tradition of making excellent varieties, and as a nation they eat on average 57 lb (26 kg) per person per year.

DRAFT & LEATHER

The economic contribution of draft cattle is difficult to gauge as they are more frequently utilized in developing countries. In Uganda, for example, each household with an ox made $245 profit per ox—essential income in rural areas—and this does not even include the effort saved in manual work.

The leather industry is worth around $100 billion a year globally. Until recently China dominated the leather market, but trade is increasing throughout the world. Demand for leather has never surpassed the provision of raw material availability; that is, cattle do not have to be killed specifically for leather/cowhide, rather the material can be stripped as a byproduct from cattle used in the beef and dairy industries.

Above: *The leather industry is worth around $100 billion a year. Cattle skin can be worth up to 10 percent of an animal's economic value.*

Dairy Farming ✤

There are two main types of modern cattle: the taurine and the zebu, or indicus. Zebu cows generally produce less milk than the taurine breeds. Crossing a taurine with an indicus can increase milk yield for the indicus breed. While a beef cow will generally produce around 1 gallon (4 liters) of milk a day, a dairy cow can produce around 16 gallons (60 liters) during peak lactation. Dairy milk yields per cow per year have increased significantly over the last decade or so— increases that are reflected in many regions worldwide.

In terms of acres farmed, China and then Australia have the largest farms, so-called "mega farms." The largest dairy farm in the world is said to be Mudanjiang City Mega Farm, with around 100,000 cows across land nearly equaling the size of Portugal and producing over 210 gallons (800 million liters) of milk per year. Another large farm is the Almarai farm in Saudi Arabia.

Left: *Dairy farms differ throughout the world. These Ayrshire cows are out on the pasture; others live year-round in barns.*

Once a cow is no longer producing the milk yields required to be economical, around 3–4 gallons (12–15 liters) a day, her meat can be used in the food chain. Typically, female calves are maintained within the herd to become milking cows, and males are slaughtered for veal or to be feeders (weaned cattle put on feedlots in order to gain weight prior to slaughter), while a few are kept as breeding bulls.

THE COST OF MILK

Milk comes at a cost. An average lactating cow must eat around 110–120 lb (50–54 kg) of food a day—normally a mixture of hay, grain, silage, proteins including soybean, plus vitamins and minerals. Getting the right balance of energy intake and the essential nutrients is critical and requires specialist knowledge. Companies invest vast amounts of money and time researching bovine feed formulations.

Milk production decreases if the cow is stressed, in pain or discomfort, or diseased. Blood flow to the udders is also increased by up to 30 percent when a cow is lying down, which in turn increases milk yield. Dairy cattle are especially affected by lameness and mastitis, which can rapidly decrease milk quantity and quality. Mastitis is a mammary gland infection that most mammals can get. It can be lethal, produce pus in the milk, and treatment usually requires administration of antibiotics. With 30–70 percent of cows suffering from mastitis in typical herds it is a serious problem. The milk cannot be used for human consumption during treatment, therefore it is a concern not only for animal health but also for maintaining financial viability and profits.

Situated in the desert, a third of its land is sheds for 46,000 cows. Saudi Arabian herds frequently reach the highest milk yields in the world. Average herd sizes have increased worldwide as higher-intensity farms tend to be more cost effective. Average herd sizes are around 375 in New Zealand, 234 in the USA (varies greatly from state to state), 220 in Australia, and 100 in the UK, but in developing countries the sizes tend to be much smaller.

Typically, the most prolific milk producers are breeds including Holstein, Red and White based upon the Holstein, Ayrshire, Jersey, Brown Swiss, Guernsey, and Milking Shorthorn. Cows produce milk after giving birth and yields peak between 60 and 100 days after calving, generally being at their highest for the following 10 months. At this point she is "dried off" (the farmer stops milking the cow so that her body, especially the udders, can prepare for the next lactation) a couple of months before giving birth again, when milk yields increase again.

COMPOSITION & TECHNIQUES

Milk is a good source of carbohydrates. The main carbohydrate in bovine milk is lactose, but this can cause problems for some people and indeed in dairy cattle. It is a sugar; that is, a disaccharide containing glucose and galactose. This process can deplete levels of blood glucose in the cow, which can create an acid environment.

Milking techniques have advanced greatly over the last few decades. Some smallholdings still milk cows by hand, but most larger herds require milking machines. Four teat cups are attached by hand, and a combination of vacuum suction and normal air pressure are used.

Above: While some owners may milk their cows by hand, milking machines and automated milking systems have shown increases in milk yield per cow.

Below: Cow's milk differs from breed to breed and even between individuals. The milk comprises differing percentages of components in comparison to other species, too.

BOVINE MILK

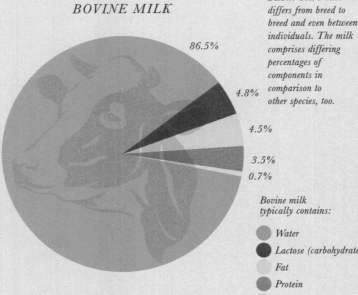

86.5%

4.8%

4.5%

3.5%

0.7%

Bovine milk typically contains:

- Water
- Lactose (carbohydrate)
- Fat
- Protein
- Minerals

Advantages of an automated milking system include:
- Cost effective for large herds
- Can reduce manual labor for farm workers
- Increased milking frequency and, therefore, comfort for the cows
- Increased milk yield
- Provides detailed individual and herd management information
- Highly developed robotic machines can attach the teats, analyze the milk, enable basic health checks of the cow, and record vital information about milk production and quality.

Disadvantages include:
- High initial costs to buy equipment
- Unsuitable to pasture systems (indoor housing is required)
- Increased electricity usage
- Concerns about lower milk quality
- Can lead to reduced interaction time with the farmer.

MADE FROM MILK

Milk production is important not only to create fresh milk but it can also be utilized in a number of ways to produce longer lasting milk, cheese, butter, spreads, and yogurts. Rennin is an important enzyme, not found in milk but naturally occurring in the gut of young mammals. This enzyme causes milk to become semi-solid, a process called curdling. This also happens to milk if left in the open air; after a few days the milk "turns sour," but it is a very natural process in calves and, of course, is used intentionally to make cheese. Once the solid parts form, the original milk grows more transparent and becomes known as whey, which is often used in protein shakes and drinks. These days enzymes, acids, and even salts are used in the curdling process. Creating cheese has become a skilled art and science, with differing processes needed for cream cheese, cottage cheese, and various other types.

Block butter is entirely made from milk fats; spreadable butter is butter mixed with oil. Dairy spreads often include milk, but functional dairy spreads may also include ingredients to help lower cholesterol levels, for example. In addition, dairy products may also be processed to create fats for cooking and baking. Bacteria are added to milk to cause a process called fermentation, which results in yogurt, the range and types of which have increased greatly over the years.

Above: *Dietary factors have a direct impact on the quality and quantity of milk produced. From fibers to fat, minerals to energy, the entire feed must be optimized.*

Left: *During the cheese-making process rennet is added to cultured milk. The proteins form into a curd (solid) which can then be separated from the whey (liquid).*

Beef Farming ✎

There is a careful balance on any beef farm between getting enough good quality meat and the costs of feeding and keeping the herd. In general, beef from cattle aged 12–18 months old is more tender; under the age of 2 years there is little difference in the meat tenderness. Beef breeds are usually mated naturally, in comparison to dairy herds, which tend to use artificial insemination. Traditionally, beef-breed calves will stay with their mothers for longer than dairy cattle, being weaned at 5–10 months old. This depends on their body condition, their mother's condition, food availability, and other factors such as the environment.

Generally, beef cattle are fattened in pastures or feedlots/feed yards. Some of the more intensive industries use beef units, where cattle have little or no access to pastures. The stages of calf rearing, growing, and finishing/fattening are frequently undertaken at different farms which specialize in each process. Eating between 1.4–4 percent of their bodyweight in feed each day, needing a stress-free, disease-free, safe environment takes a lot of maintenance. Once the cattle reach the desired weight and body condition they are sent for slaughter.

Below: *Pastures and feed lots, such as this example in Columbus, Colorado, are used to promote weight gain, producing quality beef from cattle.*

MEAT QUALITY

A type of tissue within the muscles called collagen becomes more mature as the animal ages. In younger animals, the collagen forms gelatin, but as the collagen matures, the crosslinks between these fibers maintain structure. Animals that are grain fed finishers undergo rapid growth, meaning they are ready for slaughter quicker. Studies have indicated that their collagen is immature, and, therefore the meat is more tender.

The amount of intramuscular fat, or marbling, also indicates beef quality. Again, differing cuts will vary. Breeds such as Angus and Wagyu generally have higher levels of marbling when compared to Charolais, for example. Cereal grains change the fat color from yellow to white, which increases the market value. Veal meat famously has extremely low fat levels, as young animals have not had the time to convert feed into fat. Overall, beef quality is assessed by looking at the maturity of the meat alongside the amount and distribution of marbling. Features such as firmness, texture, color, and taste of the meat will also be influenced by these factors.

As food type, additives such as estrogens, sex, genetics, and breed all play a role in meat quality, there is no exact age when animals are slaughtered for meat, but generally 12–24 months is considered usual across many countries. Older animals will be used for ground beef and cheaper products. In general, animals for slaughter are sent to specialized slaughterhouses, with the exception of ill or injured animals that cannot travel; these are often euthanized on site. Slaughterhouses usually stun the cattle prior to slaughter. Stress can cause undesired tastes in many meats and is obviously not beneficial to the animals or their handlers. Therefore, most slaughterhouses follow preslaughter handling, stunning, and slaughtering guidelines and laws. Stunning uses various mechanical (a bolt is most commonly used), electrical, and carbon dioxide gas methods. The meat is then inspected to ensure it is fit for consumption.

Below: *Marbling qualities in beef cuts vary depending on cattle age, breed, food quality and type, where the meat is from on the body, and even exercise levels of the animal.*

Muscle

Intramuscular fat

BEEF-RELATED PRODUCTS

In addition to providing dairy and beef products directly, cattle also provide us with:

- Food ingredients such as gelatin, used in a wide variety of candies and other recipes

- Beef fat, commonly used as a food ingredient, also in products such as soap

- Bovine enzymes in laundry detergent

- Glue, from hooves and horns

- Ointments and medicines, such as some antirejection drugs used for human organ transplants

- Pet food for dogs, cats, and other animals

- Delicate bone china, made from bones

Byproducts can even be found in toilet paper, candles, crayons, ink, beauty products such as shaving cream and deodorant, antifreeze, and vehicle tires.

Working Cattle & General Uses

Whether we live in the city or the countryside, many of us will have seen cattle working. Children may first see a real cow in the milking parlor on a school visit, or in a field or barn, pulling a cart on the road, or pulling a plow in the field. Cattle are useful for so much more than just food products.

PULLING THE PLOW

Although seeing cattle pulling plows or carts may not be common in all countries, in others it is an important part of everyday life and the economy. These cattle are called draft cattle or oxen. Draft cattle must be well fed and start training from around 2–3 years old in order to develop the musculature, temperament, and stamina needed for work. Once they are 5 years old and weigh 880 lb (400 kg), they will likely be able to pull a plow. Males with a docile nature are chosen, which usually means they have been castrated. Trained individuals will work well to verbal commands or a stick/whip/goad to reinforce directions. Dairy cattle tend to be unsuitable for this type of work due to their body shape, but those reared for meat, or mixed dairy and meat, are. The breed is also important, as it must be suited to the environment and weather of

the region, and it must have a straight back in order to endure the work without injury. Getting the correct harness/yoke for cart and plow pulling is key to ensuring that animals are not injured. Draft cattle can pull more weight than horses—up to two times their bodyweight for fully trained mature males. Cattle are calmer than horses, but, on the whole, are slower.

Below: *Long before tractors existed, plows were originally pulled by people, then oxen, and then mules and horses. These traditional methods are still used worldwide.*

RIDING OXEN

With cattle pulling plows and carts, it would seem natural that people would ride them; however, it is increasingly unusual to see ox or ox-back riding. Frequently called a "riding steer," oxen are usually trained as a calf using a variety of methods including leading, using a handheld clicker, or whistling; the animal learns to respond to sounds, rewards, pressure, and voice commands. Saddles, reins, and a bridle are used, which are designed to fit the animal, in a manner similar to that seen with horses. Instead of a bit in the mouth, however, as seen with horses, a nose ring (which stimulates the nociceptors if pulled) is used more frequently in ox riding. As with draft work, commands are generally vocalized by the rider and reinforced with a whip/stick. Cattle can be more uncomfortable to ride than horses as they are wider and generally slower, too. The difficulty in shoeing and getting specialist riding equipment, combined with the lack of speed and comfort, make ox riding far less popular than the use of equines. Without careful training, steers and bulls will buck and grow violent when ridden, which is used for sport and entertainment at rodeo events. In contrast, well-trained cattle can undertake walking, running, and even jumping courses.

Above: *Cattle are intelligent animals and are amenable to training. They may not be the quickest or most comfortable animal to ride but they are strong.*

LEATHER & HIDES

Used throughout the world for millennia, cowhides include the skin and hair, which is not bleached. In order to maintain its condition, the hide is salted and then processed in a tannery, where it may be dyed to achieve different patterns and colors. The hides have been used as bedding for people and pets, floor rugs, decorative pieces for the home, and even car seat covers. Tribal cultures have used them to make shields, skirts, loin cloths, headdresses, and body decorations.

Leather (including suede) is frequently developed from cowhides and is also tanned, but includes a hair-removal process. Flexible but extremely durable, leather is used extensively for clothing and accessories such as handbags and suitcases, shoes, horse saddles, furniture, sporting equipment—especially balls and boxing equipment—even book covers. Leather has existed for over 5,000 years, during which time tanning techniques have changed greatly; now, the leather may be oiled to waterproof it further. Although most animals can be used to produce leather, around 65–66 percent is retrieved from cattle generally as a byproduct of the beef industry. Vegans and Hindus will generally avoid cowhide and leather products but may use modern leather alternatives or even leather created from cell cultures in laboratories.

Below: With more than 9,000 tanneries worldwide producing an estimated 23 billion square feet (2 billion square meters) of leather, this material used by ancient civilizations is still popular.

PUTTING DUNG TO USE

Cattle create dung, and lots of it. They also produce methane. Cow dung (also known as fecal matter, cowpats, cow pies, cow chips, or manure) consists mainly of undigested plant material and minerals. Ranging from green to brown and black colors, it has been used creatively over the years. Traditionally, manure is spread on fields as a fertilizer to enrich the soil. Dung can also be used as a fuel, and the modern renewable energy source is called biogas. The material is digested by microorganisms including bacteria to produce methane, hydrogen, and/or carbon monoxide. Energy is released when oxygen is introduced, and this can be converted to heat or electricity. As dung is rich in methane it is a good material to use.

The more traditional method of using dung as a fuel is by drying it out, often in the sun, and burning it. Burning dung also has the beneficial side effect of deterring mosquitos. In some places, particularly India, the dung is mixed with water and sprayed around houses to deter insects. Dung is also great for insulating homes and is used in "mud bricks" by adding it to the loam, sand, mud, and water. Mud bricks have been used for buildings for over 8,000 years.

Dung is also used for cultural and religious reasons. It is burned in Hindu ritual fires (yajna) and as an ingredient of panchagavya/panchakavyam (a mixture used in Hindu rituals), which also contains milk, urine, curd, ghee (a type of butter), and other noncattle products. Once fermented, this mixture can be used as a medicine (often called "cowpathy"), fertilizer, pesticide, and in cosmetic products.

Dung fertilizer

Dung as fuel

Cattle in Ancient Society ❧

Although cattle have been around for 2 million years, they have not always been farmed. We do not know when cheeses were first made, but the oldest evidence shows that cheese was likely to have been made in Poland in 5,500 BCE, and an archeological site at Pokrovnik, Croatia, shows remnants from 7,000-year-old cheese. The fact that it was placed in Egyptian tombs dating back 3,200 years ago, and beside the bodies of mummies buried in China's Taklamakan Desert some 3,800 years ago, shows that cheese must have played an important role even in those days. It is likely that many of these first cheeses originated from sheep and goats, but today the vast majority comes from cows' milk.

A cave excavation in Armenia revealed a cowhide shoe which is 5,500 years old, and similar leather products have been uncovered in Stonehenge, England, from around 5,000 years ago, in the Pyramids of Giza 4,500 years ago, and even with a leather manuscript from ancient Egypt. It is even possible that a 40,000-year-old Neanderthal deer-bone tool found in France may have been used to soften and shine leather—well before the existence of modern cattle. Leather was well used by the ancient Greeks (twelfth to ninth centuries BCE), with Homer mentioning it in his writings. The Romans not only fed their armies on cattle products but also used leather to dress their soldiers.

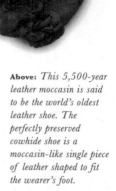

Above: *This 5,500-year leather moccasin is said to be the world's oldest leather shoe. The perfectly preserved cowhide shoe is a moccasin-like single piece of leather shaped to fit the wearer's foot.*

Left: *Roman soldiers were often clothed in leather. The so called* pteruges, *often strips of leather attached to armor, protected the shoulders and hips.*

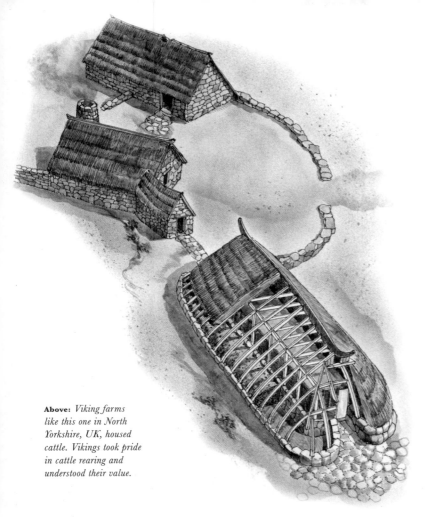

Above: *Viking farms like this one in North Yorkshire, UK, housed cattle. Vikings took pride in cattle rearing and understood their value.*

THE VIKING VALUE

Even the Vikings were keeping cattle from the eighth to eleventh centuries. DNA evidence from Ireland, bones found in Greenland, and excavations of farms across Norway, Iceland, and many other countries have all helped to piece together the farming habits of these people. Due to the harsh climates in which the Vikings lived, these cattle were often kept indoors and fed off crops during the cold winter months, then taken out to pasture during the summer. In the early days, the Norse people would live in a longhouse with their animals alongside them in stalls within the building. As farming methods developed, they moved the animals into outhouses. The Vikings valued dairy products but did also eat meat from the cattle. Fresh dairy was eaten all year round, but skyr (curds) and sour milk were stored to help feed the people through the cold winters, when milk production was low. Bulls were sacrificed to the gods, and oxen (part of the *Bos taurus* subfamily) were used to pull plows and sleighs.

The value of cattle was very high: the word for money and cattle was the same for these historic people. The farms were run by the women and children, as frequently the men would go on raids during the summer months; therefore, farming was very much a family affair. We also understand the importance of cattle when looking at the laws and expectations of society. In Iceland, people paid a tax toward a communal insurance called *hreppr*. If a farmer lost buildings or cattle they could claim on this insurance to help cover the loss. In short, cattle were highly valued and an essential part of farming and the community.

Below: *Vikings generally wore leather boots or shoes, tied with a leather strap. This tenth-century example was discovered in the town of York, England.*

Cattle in Religion ✥

HINDUISM

Cows are thought to be sacred by those who follow Hinduism, and are celebrated for their strength and gentle natures. The cow plays a major role in the celebration of several festivals: Mattu Pongal celebrates the work cows do for agriculture, and Gopashtami is dedicated to Krishna and cows. The Hindu goddess Bhoomi is often depicted as a cow, and the goddess Kamadhenu (also called Surabhi, Shurbhi, or Gayatri), the divine cow, is the mother of cows. Hinduism goes further than simply revering these animals, though. India has a large number of followers and also has the largest number of cattle in the world. Once cattle are older or unwell, they are cared for in Gaushalas (protective shelters), of which there are over 3,000 in India alone.

Various other members of the bovids are also revered as sacred within Hinduism. The close relative the nilgai (an antelope) is associated with the cow and is, therefore, also sacred. Indeed, despite being an antelope, the clear physical similarities between cattle and nilgai help us to understand why this association has been made. Hinduism promotes a beef-free diet but permits the use of dairy products; many followers are strict vegetarians. The religion is the third largest in the world, with around 15 percent of people globally following its teachings; it is also thought to be the oldest religion still active today, at over 4,000 years old.

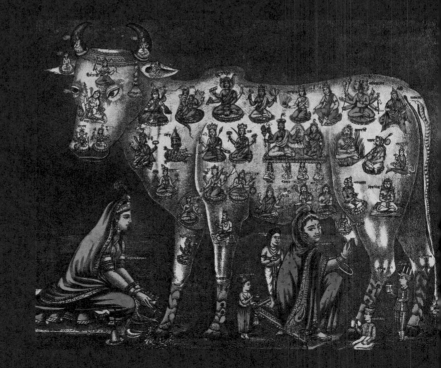

Below: *In some cultures cattle are not just appreciated, they are often worshiped and considered sacred. The Hindu goddess Bhoomi is often depicted as a cow.*

OTHER RELIGIONS

The range of ethics surrounding the use of cattle is sometimes based on religious or philosophical grounds for other people, too. Christians and Jews can seek to follow a vegetarian diet based on their religion but may choose not to. The Hebrew bible discusses a number of sacrificial cows, and unblemished red cows were important in ancient Jewish rituals. Jewish and Islamic people believe in rituals around cattle, and while many will eat beef, it should be slaughtered in a specific manner. Meat will often be called "kosher" or "halal" if it adheres to these rules. The importance of cattle is very clear in many of these religions, featuring often in the writings of the Bible, the Koran, and other religious texts.

Many religions teach abstinence of beef and/or dairy and other cattle products. Jainism promotes a vegan lifestyle that rejects the use of cattle products across the board. Buddhists are mostly vegetarian and some followers believe that cattle are reborn humans. Although Buddha himself ate meat, it is not clear whether he ate cattle, and certainly over the years vegetarianism has become an important aspect for many believers. The Chinese religion of Taoism also follows a vegetarian lifestyle, and China in general has somewhat of a taboo over eating beef. Emperors have banned it and it is rarely used in Chinese medicine.

Ancient religions often portrayed the bull as a great, strong beast. Ancient Egyptians revered the bull for its strength and virility, and often hunted and sacrificed the beasts. Meanwhile, the cow was revered for its fertility and used to represent motherhood—the cow goddess, Mehet-Weret, was said to have given birth to the Sun. The Celts, ancient Greeks, and Vikings placed cattle at a godlike level and in some cases believed that the gods kept cattle. Cattle played important roles in myths and stories.

Cattle in Society & Culture ❧

Cattle have strong influences on lives throughout the world, and have a great impact on society and culture, as well as religion. Today their reach and influence extends beyond the dairy, beef, and draft-work industries—they are part of our communities, in petting zoos, smallholdings or city farms, and on ranches. We see them at country shows, or even at one of the few remaining events such as bull runs and bullfights. Increasingly, people are even keeping cattle as pets.

Cattle are much loved and revered around the world. Our fascination with them extends beyond farm and field. Their images feature as national emblems, sports teams mascots, on television and in movies, and across the Internet. They are written about in literature, portrayed in art, and shown in museums. From advertising campaigns through to the celebrities and world leaders who keep them, cattle play a significant part in society.

CULTURAL APPEARANCES

Our awareness of cattle may start with the nursery rhymes, songs, and bedtime stories we hear as children: a cow jumping over the moon, joining in with the "moo"

when singing "Old MacDonald had a farm." As children grow older they may hear the Greek myths of Zeus turning Io into a heifer, or Europa being seduced by the god Zeus in the form of a bull; the English fairy-tale of Jack selling his dairy cow for magic beans which grow into a beanstalk; the Russian folktale about The Dun Cow; or the Himalayan story of a black cow.

Above: *Cattle have long featured in folk and fairy-tales and nursery rhymes. Jack and the Beanstalk featured a cow which was sold for magic beans.*

Cattle appear as television and film favorites and company emblems, for children and adults alike, including the "laughing cow" logo and advertisement for a popular cheese spread since 1921, and the famous bulls that make up the internationally recognized Red Bull logo. Mascots for many sporting events and teams include cows and bulls. The bull features for sports teams such as R.B. Leipzig, Germany; Deportes Savio, Honduras; Tex Hooper, the bull for F.C. Dallas; and the Chicago Bulls basketball team.

From Bollywood to Hollywood, you are never far away from hearing about or seeing bovines. Some famous cattle in TV and movies include Clarabelle, one of Minnie Mouse's closest friends; Annabelle, from the Christmas movie *Annabelle's Wish*, and the animated comedy series "Cow and Chicken." From cattle drives and western movies to romances set on ranches, cows have their role to play.

Ten flags around the world incorporate the animal, including the flags for Andorra; Caso in Spain; Kansas and Vermont in the USA; Mecklenburg-Western Pomerania in Germany; Adelaide in Australia; Sucre in Colombia; Uri in Switzerland; Moldova; and even the presidential flag of Iceland.

CAN COWS ACTUALLY JUMP?

Moon-jumping by cows is, of course, (presently) impossible, despite the nursery rhyme stating otherwise. Cows are not well known for their jumping abilities; while they can jump fences, they tend not to. As for moon-jumping, gravity works to keep cows on the planet; cows need oxygen to breath, and they would burn up in Earth's atmosphere if they attempted to jump to the moon or back. Our cow would need a great deal of jumping force and velocity if it did jump, as well as needing to increase its muscle mass immensely. Even the muscled Belgian Blue is not in the running here. While many animals have traveled to space in spacecraft, including mice, dogs, and monkeys, so far a cow has not been one of them.

CATTLE IN THE ARTS

From ancient tomb and cave markings to modern paintings and sculptures, artists throughout history have been enthraled with these great mammals. Leonardo Da Vinci sketched out the anatomy of the uterus of a pregnant cow, and the German expressionist Franz Marc is well known for his brightly colored, animal-focused creative pieces, which include *The Yellow Cow* (1911). American artist Andy Warhol created cow wallpaper and pop-art pieces based on the cow called the Cow Series. More recently, Turner Prize-winning British artist Damien Hirst caused controversy and interest with his piece titled *Mother and Child Divided*, which is a cow and calf dissected and preserved. Another of his bovine pieces was created from a calf in formaldehyde, with 18-carat gold horns and hooves, and sold for £10.3 million in 2008. Hirst's work of a cow and bull rotting was even banned amid fears of inducing vomiting in the audiences.

Sculptures of cattle are also very popular. Italian-American artist Arturo Di Modica created *Charging Bull*, which symbolizes Wall Street and commerce. The CowParade is an international art project that places artist-decorated cow sculptures across the world. The parade has become so successful that many towns and cities have copied the idea and created their own take on the statues. With their bright colors and unique appearances, they provide fun and interest to urban streets and parks, in addition to being popular for social media photographs.

Entire museums are dedicated to cattle, from the Milchmuhseums (Dairy Museum) in Germany, the Cattle Museum in Japan, the Museum of the History of Cattle in Finland, to the Cattle Raisers Museum in Texas. It is well worth seeing if your local museum has highlighted the role of cattle in your area, or checking whether you are going on vacation near one of the many specifically bovine-themed museums.

Above: *Cattle are popular themes in art and culture. This painting titled* Fighting Cows *was created in 1911 by German artist Franz Marc.*

Below: *Arturo Di Modica created* The Raging Bull *following the 1987 stock market crash; it is often called the Wall Street Bull.*

FAMOUS CATTLE & OWNERS

Celebrities across the world have not limited their appearances with cattle to the photographic lens and film—some stars keep their own cattle. Martha Stewart keeps cattle alongside her other farm animals and vegetable plots. Nicole Kidman has kept Black Angus cows alongside alpacas on her Australian farm.

You have probably heard about the First Cat and the First Dog of the White House, but did you expect to hear about a First Cow? President Taft (1857–1930) had to get milk from his Holstein Pauline Wayne when his First Cow, named Mooly Wooly, died after eating too many oats, much to the disgust of *The Washington Post*, which reported on the death. Pauline, or "Miss Wayne," as she was often called, served the White House well, and even arrived pregnant, giving birth to a healthy bull. Pauline was a star, and lives on in history as being the last official White House cow, sadly; but many went before her and people loved hearing about and seeing the First Cows. Other famous First Cows included the Durham cow named Sukey, belonging to President William Henry Harrison (1773–1841), and the Jersey cows of President Rutherford B. Hayes (1822–1893). More recently, while President George W. Bush did not keep a cow at the White House, Ofelia the Longhorn lived on his family ranch in Texas.

Presidents of the USA are not the only heads of state to own cattle; the British royal family have long kept herds. Queen Elizabeth II keeps her 100 Highland cattle at Balmoral Castle, Scotland, an active farm that ensures the royal family get top-quality, home-grown beef. The queen also owns a stunning dairy. Built in 1848, the Royal Dairy is not only a marvel to look at but has also supplied cream and dairy products to the royals for many years. The queen's herd of 165 Jersey cows descended from Pretty Polly, a cow gifted to Queen Victoria in 1871. The dairy itself was designed by Victoria's beloved husband, Prince Albert, who encouraged agriculture and innovation. In keeping with his ideals, the modern-day herd are milked by robots and the cows can lie on waterbeds. Her Majesty insists on keeping the highest welfare standards for her animals.

Zulu King Goodwill Zwelithini and his herd made headline news across the world when rustlers stole his prize-winning animals. It was reported that over 7,000 of his cattle have been stolen over the years from one breeding station alone. It is well known that the king promotes agriculture, and even introduced the Boran breed to subsistence and communal farmers to enhance the genetic pool of the indigenous cows.

Below: *A young Queen Elizabeth II and Prince Philip, the Duke of Edinburgh, walk through pastures with their Highland cattle in the grounds of their Scottish home, Balmoral Castle.*

Wild vs. Domesticated

There are large differences between truly wild cattle, feral cattle, and domesticated cattle—and no classification guarantees tameness. Wild animals will not have been domesticated in any way, nor selectively bred; their only interaction with people is accidental. Feral animals are those that have been in captivity and/or domesticated but then escaped and now live in the wild. Domesticated cattle (the word "domesticated" derives from the Latin for "belonging to the house") usually live on farms or ranches, in zoos, fields, or on managed land. They are managed, which means their food, water, and health needs are provided for by an owner.

There are many cases of feral herds or individuals living throughout the world—India has vast numbers of feral cattle.

Not owned by an individual or company, they have usually come from domesticated breeds. They often rely on humans for food and water, and many receive veterinary care. The same can be said for small herds of cattle in Africa which have escaped from farms. The cattle tend to be smaller as they are less well fed, and selective breeding is reduced as the generations pass. Not every feral herd is welcomed, though. A herd of around 150 cattle in California caused great debate when they started being aggressive toward people visiting the Mojave Preserve and Sand to Snow National Monument. The animals caused damage to the monument and reserve, and polluted the water.

Below: *Today there are two separate herds of bison in Yellowstone Park. They were the last remaining free-ranging bison herd in the USA.*

THE LAST OF THE WILD?

The last truly wild cattle in Europe are thought to have been the aurochs. Other cattle such as the gaur (*Bos gaurus*) in India, buffalo, and bison still have wild herds, but true wild *Bos taurus* are gone. The Wild White Cattle of Chillingham in the UK are said to be the last wild cattle in the world. Records date back to 1862, when Charles Darwin encouraged the idea. The herd have some contact with people, with occasional hay put down when the animals are starving and carcasses removed to be buried. This means they are managed and not absolutely wild. The animals receive no veterinary care, are highly inbred, and can be extremely aggressive toward people. Bulls reach around 660 lb (300 kg) with females weighing in at 44 lb (20 kg) lighter. With around 100 individuals, the Wild White is rarer than the panda; weak calves are left to die, while six to eight bulls sire regularly. No outside cattle have ever entered the herd, but a few Wild

Whites have been removed for breeding elsewhere. The Aleutian are feral cattle found in Alaska which have avoided full domestication, as have many other herds around the world.

Some rearing methods are resulting in cattle becoming less managed over the years. Management of ancient and damaged land is a way in which cattle are increasingly being promoted to benefit the land and the climate. These techniques have also been suggested as a method of reversing or reducing desertification, the process by which fertile land becomes desert due to inappropriate agricultural use, deforestation, or drought. The theory is that cattle grazing and manure help develop the land, promoting plant and animal biodiversity. Countries ranging from the USA through to Scandinavia are increasingly employing these techniques. Methods are constantly critiqued, debated, and researched in order to better inform for the future.

Above left: *The Wild White Cattle of Chillingham are often referred to as the last wild* Bos taurus. *Nearly extinct, they are monitored carefully but have little human interaction.*

Above: *Although* Bos taurus *are not usually wild, other cattle such as the Indian gaurs (pictured), buffalo, and bison still have wild herds today.*

Modern Interactions & Behavior With Humans ✑

Cattle can live in or next to people's homes, in vast farm management systems, or in feral situations. Cattle behavior and interactions therefore depend on the level and type of contact with people. Whether they are part of our food production or a pet, the way we interact with cattle and our expectations of their welfare needs are advancing over time.

In general, people are becoming more informed about farming and slaughtering methods, and our interactions with cattle are shifting. As farms become more intensive, with cattle being housed indoors (which often results in higher milk yields and has a lower cost to the environment), people appreciate seeing free-range cattle in fields. And as urban areas grow, people are also more interested in experiencing interactions with tame cattle.

THE IMPACT OF MODERN FARMING

In the dairy industry, mechanical milking machines, robotic milking technology, and indoor housing have become the norm, meaning less hands-on time for milking and herding cattle. In developed countries, this often results in fewer people around cattle. While this may reduce cattle stress, it may also reduce how easy cattle are to handle, and impact negatively on their behavior around people. Further, while automatic feeders help farmers control food intake and maintain good health, they also remove the strong and positive association between people and food. Cattle taming and good behavior is often achieved using food as a motive; this association decreases if food is given by a machine. Since the 1940s, farming methods in many places have meant that the time spent with each animal has decreased. This may increase the fear shown by cattle toward people, make animals more

Below: *This cow in India, taking food from people traveling in a train, shows how cattle can learn and adapt. People and cattle often establish close bonds.*

Above: *Feeding cattle can be hard work. The sheer weight of food they need to consume each day means heavy haulage for farm workers. Whether helped by machines or having to grow and distribute the food by manual labor, it is a time-consuming process.*

difficult to handle, increase the risk of injury for both parties, and even cause stress and health problems for the animals.

However, other modern techniques increase human interactions. For example, artificial insemination requires cattle to be relatively tame, and vaccinations, pregnancy and parturition care, and health testing are times when close contact is maintained. So, although farming methods are evolving, the human–cattle bond still exists and is an essential part of the industry. Although no studies prove a genuine cattle attachment

to humans, there is no doubt that individuals recognize particular people and behave differently depending on how that person has treated them in the past.

Environmental Impact ❧

There are increasing concerns over cattle and their impact on the environment, as well as over the use of medicines and drugs in cattle rearing. The United Nations has highlighted that cattle are a major stressor on ecosystems and the planet as a whole. Livestock farming is also the leading source of water pollution and cause of loss of biodiversity.

GREENHOUSE GASES

The carbon footprint has become an important concept to many consumers and governments throughout the world. The cattle industry and many farms are now concerned with balancing the economies of cattle and milk production with understanding and addressing the environmental impact. Data from 53 countries has shown that carbon dioxide (chemical formula CO_2) emissions range from 90 kg CO_2 to more than 1,000 kg

Below: Reducing carbon footprint is important not only for the environment but also in making farming in general more viable.

CO_2 per 100 kg milk. Carbon footprints from large, high-yield farms in developed countries tended to have smaller carbon footprints in comparison to low-yield systems in developing countries.

Methane (chemical formula CH_4) is also a greenhouse gas, like CO_2, but the negative effects of methane are likely to be 23 times higher than CO_2. On average, 100 kg of methane is produced for each cow per year, equivalent to 2,300 kg CO_2. Dairy cattle in the USA and Canada produce an average of 158.7 kg of methane per cow per year (58.8 kg in the meat industry), whereas in Africa that is reduced to 77.2 kg (31 kg in the meat industry). An important factor to remember is that their average milk production yields also differ greatly: U.S. and Canadian stock produce an average of 8,400 kg of milk per animal per year, compared to just 475 kg in African areas.

Contrary to popular belief, livestock are not the dominant source of global methane emissions. Livestock across the world (including cattle) account for around one fifth of global methane emissions. Production of petroleum and natural gases, coal mining, landfill waste from homes and businesses, and even natural wetlands all produce methane. Smaller amounts are produced by the oceans, volcanoes, and wildfires. Although CO_2 is naturally present in the atmosphere, human activities change the amounts. Fossil fuel combustion for heating, electricity, and transport is a key emitter, as are industrial processes. Nitrous oxide (chemical formula N_2O) is another greenhouse gas produced by human activities such as agriculture, fuel combustion, industry, and from waste; 65 percent of atmospheric N_2O comes

from the livestock industry presently. In relation to cattle, nitrous oxide production is predominantly due to manure management, fertilizers, and growing their food.

Manure management is important in relation to N_2O production and cattle. The amount of methane, CO_2, and other gases produced per animal is based on a number of factors including which manure management systems are used; feed intake quantity and quality; milk production quantity; how much the cattle are growing and, therefore, how much energy they need; amount of draft work and foraging; and pregnancy status. Manure production is a serious matter when discussing the environment. Dairy cattle produce roughly 4,860 lb (2,205 kg) manure dry matter per animal per year, and meat/other cattle produce 3,330 lb (1,510 kg). Manure management systems have advanced greatly over the years and are now part of the solution to reducing environmental impact. Management types range from being deposited on pasture, short-term pit storage, long-term anerobic lagoon treatment, use in building houses, and burning, with some methods having lower emissions than others.

Above: *Manure can be a useful byproduct as fertilizer, or even to build homes, create energy, and for heating.*

DEFORESTATION

When looking at overall environmental cost, it's worth considering that cattle eat plants—those grown for them and the byproducts from plants grown for people which are not consumed by people. These include Brewers grains, byproducts from oilseed crops, and sweet-corn cannery waste. In addition, farmland reduces biodiversity in comparison to forests or completely wild land, although it usually has more biodiversity than urban land. Much of the global deforestation is driven by cattle ranchers. Brazil has huge numbers of cattle, and over 80 percent of the Amazonian deforestation alone has been to make space for the cattle industry, over 280,000 square miles (450,000 square kilometers) to date. Deforestation also adds to the release of carbon into the atmosphere each year, in addition to reducing biodiversity, increasing fire risks, affecting the flow of rivers, and causing soil erosion, among other environmental impacts.

TOWARD SUSTAINABILITY

Farmers, the industry as a whole, the public, and governments are highly aware of the impact of cattle on the environment, and there is a drive to reduce the negative impacts worldwide. Monitoring of land use, government and nongovernment organization interventions, and even financial implications have been set out in order to reduce deforestation, for example, in favor of more sustainable farming. Research is constantly being undertaken in order to reduce emissions, make cattle more effective and therefore less of an impact on the environment, and toward developing more sustainable farming and management methods.

Precision Livestock Farming aims to reduce the use and production of antibiotics, phosphorus, heavy metals, nitrates, ammonia, and greenhouse gases. In tandem, precision livestock farming aims to ensure animal welfare, health, and reproduction rates, and maintain or increase production and economic viability. Many techniques are applied, including monitoring each animal

Above: *The cattle grazing in the Amazon region must be grazed in a sustainable manner. With land use at a premium, deforestation is not always welcomed.*

carefully using technology-based systems; optimizing feed to be economically viable while environmentally friendly; the use of robotic milkers and automatic feeders that constantly evaluate the animals and products such as milk to check for infection, in order to reduce widespread antibiotic use; checking and controlling the environment using automated systems to reduce stress and maintain good conditions.

Other drives to reduce environmental impact have concentrated on the full utilization of each animal. This includes using all products from each animal. For example, reducing dairy and meat wastage in supermarkets and homes, and using each animal produced to its optimum. Organic or naturally grown cattle do mitigate some of the impact and are becoming increasingly popular in many countries, despite the higher cost to the consumer. Organic farms generally require more space for grazing and/or crop growing, do not use fertilizers or artificial pesticides, provide a grass-rich diet that is GM free, and ban the use of routine antibiotics. Milk yields can be lower and the farming more expensive, as crop loss can be higher, growth rates slower, and labor more intensive; but on the whole it is considered to be a better system for wildlife and the environment. While most animals are free range, this is not specified in all countries, therefore the labels "naturally grown" or "free range" highlight that cattle spend as much time as possible in pastures.

Below: *Some countries are able to produce a lot of free-range animals, such as these grazing cattle in New Zealand.*

Ethics & Welfare ✇

Exploiting cattle in sport, especially bullfighting and bull running, has caused much discussion and protest in relation to ethics and welfare issues. The bull in particular has been used historically for competitive runs, fights, and even gladiator games due to its size and aggression when provoked. While some regions and countries have banned these practices, others maintain the "sports" but with modifications to exclude physical injury and prevent death. Many argue that the stress caused to the animal still poses a welfare risk—as is also often the case for the people participating or spectating—and reduces the animal's quality of life. Many traditional runs or fights have been replaced with mock runs, where cattle are replaced with fast human runners. In other places, the traditional cultural sports have been replaced in their entirety, with marathons and even pub crawls taking place instead, thereby still creating a large social gathering but without the need for animals.

Below left: *These animals are in a corral for the largest live cattle auction in Argentina, Mercado de Liniers.*

Below: *Bulls in particular have long been used for entertainment and sport. Here, calf roping recreates the skills used by cowboys / animal herders to catch cattle.*

CARE OF KEPT CATTLE

Animal welfare and care is also of paramount importance for the farming industry. There is no doubt that over the years, both farming and veterinary care and medicine have advanced greatly in some countries, whereas in others more traditional techniques are still used. The same can be said for educational and conservation schemes including zoos, petting farms, city farms, and similar programs. Some believe that no animal should be kept in captivity; others believe that housing and care should be better than presently available. Balancing income, costs, time, space available, and safety of animals and people with the needs of each animal can be difficult, but, increasingly, society and professionals want to create the right environment.

Other issues debated revolve around transport between farming/zoo locations and also to the slaughterhouse, where necessary, and the environment of the slaughterhouse itself. Only animals fit enough to travel can enter the slaughterhouse—any that are not well enough (for example, severely lame or dying during calving) must be euthanized on the farm by an appropriately trained person.

The methods of slaughter are also discussed and researched at length. While not everyone may agree with the killing of animals, ensuring painless and stress-free conditions is essential. The technologies available, staffing levels, money available, and even local customs, cultures, and religions can all have an impact on the decisions made around this process. In some areas, trained veterinarians are employed to ensure that strict laws are adhered to; in other parts of the world, the local butcher, farmer, or even religious leader might undertake the task.

Below left: *Differing laws and ethics are applied worldwide to cattle transportation and the methods of slaughter.*

Below: *Petting farms and zoos help children understand animals and animal welfare. These farms are often rescue sanctuaries for animals.*

ETHICS OF CONSUMPTION

There are a number of people worldwide who do not eat dairy and/or meat products, refuse leather products, do not believe in using animals for human benefits, and may even decide not to keep pets. Some limit this to not consuming cattle products specifically, while others include all animal products and byproducts. Other people believe in plant-based or vegetarian diets based on the belief that animals should not be killed or experience pain; they might wish to see better universal welfare during the

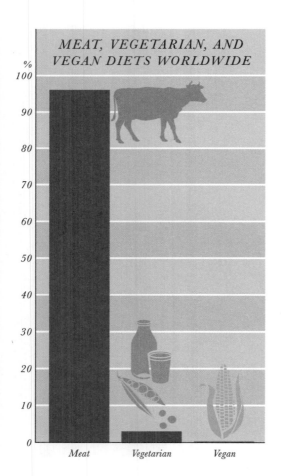

MEAT, VEGETARIAN, AND VEGAN DIETS WORLDWIDE

lifetime of animals, such as animals being free range or with better housing conditions. Some people may not want to see grain eaten by animals when it could potentially be fed to people, as food is a valuable resource, or they may be concerned about the environmental impact of cattle farming. Increasingly, there are also a number of people with allergies, or those who decide to avoid animal products for a particular health and lifestyle diet. In addition, many people do not eat beef or dairy products simply because they are too expensive or not available.

It is difficult to know how many people in the world are vegetarian/vegan, but estimates put it at around 20 percent, which includes those who make an active choice and also people who cannot obtain or afford animal products.

Left: *This chart shows the percentage of people who are meat eaters, vegetarian, and vegan worldwide. Google search trends increase yearly for the words "vegetarian" and "vegan."*

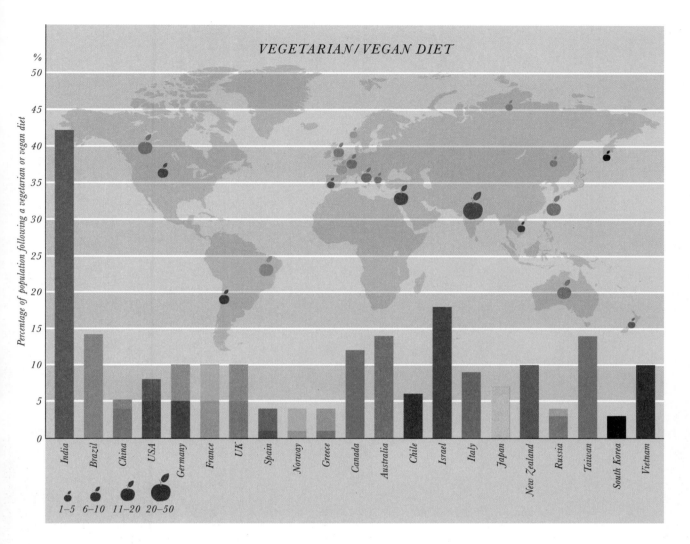

VEGETARIAN/VEGAN DIET

Percentage of population following a vegetarian or vegan diet

%
50
45
40
35
30
25
20
15
10
5
0

India | Brazil | China | USA | Germany | France | UK | Spain | Norway | Greece | Canada | Australia | Chile | Israel | Italy | Japan | New Zealand | Russia | Taiwan | South Korea | Vietnam

1–5 6–10 11–20 20–50

Some people may prefer to eat meat and dairy products but demand organic products. They may also prefer grass-fed, pasture-reared products, and may eat beef but reject products such as veal, as the meat is from younger cattle and traditionally veal calves were reared in small spaces. Modern farms can have very different practices in comparison to those of the past, giving consumers more choice over what they eat and how it has been reared.

There are many ethical and moral views on cattle and their care, which combine with the many laws and protective measures applied in different parts of the world. Whichever side of the debate resonates the most, those who own cattle are more likely to make better economic returns from healthy, well fed, cared-for animals, and this helps to regulate the business.

Above: *The number of people who follow vegetarian or vegan diets throughout the world changes, and is often dependent on religion, culture, and socio-economic factors.*

CHAPTER 5

A Directory of Cow Breeds

Historic Breeds ✑

With around a thousand modern breeds across the world, and a billion animals on Earth, cattle have evolved and adapted to differing environments. Many have been bred in order to produce beef/milk or as draft or sporting animals, but the basic characteristics prevail. These include the ability to remain warm or cool in varying climates, resistance to disease and parasites, being hardy by nature, and feed efficiencies. Modern farming takes advantage of historical breed characteristics and enhances them.

North & South America

When America was first discovered by the Europeans it contained no cattle. In 1493, explorer Christopher Columbus introduced now extinct Indian and European wild cattle. In the sixteenth and seventeenth centuries, more cattle were transported to Central and South America which have developed over time. At first, most of the New World herds were semi-feral and mostly unmanaged. Therefore, natural selection took over for around 450 years—between 80 and 200 generations—to produce the Longhorn. Meanwhile, back in Europe and other parts of the world, breeding programs had already started.

Indicus cattle are better adapted to living in warmer and tropical climates; therefore, despite the first cattle being the European taurine, over the centuries the zebu breeds have become more numerous in America. Taurines do still live in the USA though. The criollo breed is a good example, although it is a taurine that has adapted to more tropical climates; Colombia has a similar breed called the Costeno con Cuernos. Today, countries such as Brazil and the USA house high numbers of cattle. The USA topped the 2019 table of beef and veal producers worldwide at 14 tons (12.73 tonnes).

Above: *Criollo cattle are taurines that have adapted well to heat. They are descended from the first cattle brought to the Americas by explorer Christopher Columbus.*

oldest surviving European breed. There is also evidence that Salers from France existed 10,000 years ago, but given that they are depicted only in cave paintings (which can be difficult to decipher and/or date) and not written or archeological evidence, this is difficult to ascertain. Despite being an island, Britain also has its fair share of older breeds, such as the Sussex, which was described in written chronicles from 1066, at the time of the Norman invasion of England.

Left: *Shipping methods have changed greatly over the years. Livestock carriers now carry up to 25,000 tons of cargo, crew, provisions, and up to 20,000 cattle.*

Below: *Chianina cattle have existed for more than 2,200 years. This beef breed would have roamed the planet alongside now extinct bovines.*

Europe

Traditionally, European cattle mainly descended from taurines (from the Middle East and European domestication events), but there is evidence of indicine lines in the past, too. Much research has been carried out into where and when breeding and domestication events occurred, including looking at ancient and modern DNA. It is still thought that there were several points at which cattle were introduced into Europe. These probably consist of both natural movement of animals but also via shipping of animals to differing countries—this is certainly true in more modern times, of course.

The oldest cattle in Europe were the original aurochs, which became extinct in 1627. Today, there are nearly 500 recognized breeds residing in Europe. These include older breeds, such as the Chianina from Italy that have survived for at least 2,200 years and is possibly the

Australasia

Cattle are a relatively new species in countries such as Australia and New Zealand, with the original Auroch failing to reach these regions. The first two bulls and five cows were imported to Australia in 1887 from South Africa, and the cattle brought from Britain to New Zealand in 1814 were pivotal in the history of breeds in Australasia. The *Bos indicus* animals tended to survive the often changing and extreme temperatures, and tended to be better adapted at grazing on poorer quality lands and surviving periods of low water levels.

Asia

At the heart of cattle domestication, Asia saw the rise of taurine cattle from domestication events around 10,000 years ago in the Near East and potentially around the same time in China, and later the rise of zebu in southern Asia.

Management of cattle resulted in changes in behavior such as increased docility, changes in skull, skeleton, brain, and horn morphology, and changes in the quality of beef and milk production yields. Another way in which we can see domestication event timing is by looking at bone damage, caused by activities such as draft work. Combinations of archeological dating, morphology, genetics, historical documents, and even looking for disease effects in animals helps us to track the evolution and movement of cattle.

Today, India is especially well known for housing a very large number of cattle, despite the fact that few people eat beef in this country. Worldwide milk production in 2019 was set at around 268 billion dollars, with India, the USA, China, Pakistan, and Brazil often topping the league tables.

Africa

The first African cattle were present in Iraq, Jordan, Syria, and Israel, in the Fertile Crescent (see page 27). With similar genetics to those from 10,000 years ago in the Middle East, the original cattle were bred with the native aurochs. More modern breeds such as the Sanga date back to 1600 BCE, and today over 150 indigenous breeds have been registered or recognized. These range from *Bos indicus* to *Bos taurus*, but many are nearing extinction as they are farmed in remote farms or restricted to small areas. In addition, there are likely to be many more breeds still not officially classified. African cattle have been exported worldwide, as they are often able to deal with differing climates and environments. Animals such as the yak, gaur, banteng, and water buffalo were also domesticated in this region.

Above: *Ancestors of Watusi cattle can be traced back to 4000 BCE in Egyptian pyramid pictographs. They migrated to Africa, and modern breeds now exist worldwide.*

Left: *Shepherding and farming techniques vary greatly from feral, traditional shepherding, such as this group in Uganda, through to large intensive farms.*

Modern Breeds & Breeding ❧

Modern breeds tend to be classified as either dairy or beef, purely because in order to stay competitive within the market, particular features must be bred into the cattle and maintained. Breeders may want a hardy animal that will be resistant to disease and infection, calve without assistance, manage to survive in differing climates and even drought conditions in some areas, and cope well with their environment, including damp conditions in fields or direct sunlight in warmer countries. With careful breeding, alongside genetic testing, vaccination programs, and medical and surgical treatments, the modern cattle breeds are evolving to provide the farmer with a profitable and viable animal.

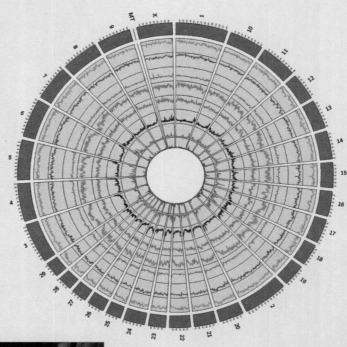

Left: *Scientists research both modern and ancient cattle using genetic techniques in the laboratory. Genetic data is often used alongside clinical, farm, and production data.*

Above: *Cattle genome projects have not only sequenced the bovine genome, but have also discovered many genes and mutations of importance for disease and breeding.*

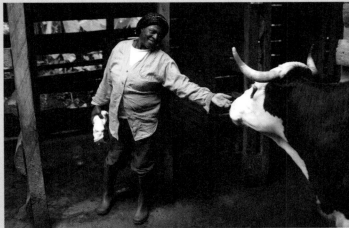

DESIRABLE TRAITS

Tameness

One key feature which has been of great importance in most breeds has been ensuring easier handling. Tameness is one factor, therefore, which has been bred into many breeds over time, facilitating the care—and slaughter—of animals. It is likely that even the original aurochs were bred based on their temperament, the calmer ones being easier to handle. Cattle have also become more tame due to being handled, herded, milked, and other human interactions. That being said, in present day farming, productivity is often more important than tameness, therefore some breeds may still be nervous and/or aggressive. Generally, it is still true that the "flight zone" is smaller in dairy compared to beef cattle. If people or other objects move into this zone, the cattle will move. The zone size around each animal changes depending on the age, sex, amount of handling, and previous experiences, but it is likely that genetic selection also plays a role.

Polling

Another consideration for modern breeders is that of polling (breeding animals without horns). Many breeds naturally have horns while others are naturally polled. Increasingly, more breeds are being bred to be polled. With ethics an ever-increasing factor in modern farming, maintaining safer handling of the animals while not having to dehorn/disbud is becoming more popular. In addition, it can be a more economically viable option, as the cost of these procedures is negated.

Hardiness

Despite brilliant advances in medicine and science, the agricultural industry is facing concerns over antibiotic resistance. The cost of treatment, preventative medicine, and general care of the herd is also of importance. Therefore, herd resistance and hardiness, plus genetic conditions and traits, are becoming priorities.

Above left: *Polled cattle, such as this bull, are often desirable due to increased safety for workers. Breeds may be polled naturally or horns can be removed.*

Above: *Dairy farmer Mary Sirri Ndikum from Cameroon, central Africa, owns several cattle. Her business delivers milk and produces yogurt.*

SURVIVAL & DIVERSITY

Extinction is a very real threat to today's cattle, despite the high numbers currently roaming the planet. Many efforts have been applied to try and prevent modern breeds from joining now extinct breeds such as the Abruzzese, Alderney, Ancient Egyptian, Fribourgeoise, and Niata cattle. Breeds such as the Scottish Shetland, the American Miniature Panda, the Sahiwal from Pakistan, and the British Vaynol, to name but a few, are endangered and have conservation plans to increase numbers. Many countries have indigenous breeds that are thriving, yet others are suffering from reduced production capacities in comparison to other breeds, low fertility levels, late maturation, high costs of management, and even loss of habitat. Advances in reproductive medicine, such as embryo transfer, surrogacy, semen storage, and artificial insemination, are essential in these situations.

Talk of Jurassic Park-style farms in order to recoup costs of keeping rare cattle has hit the media. The reality for some breeds is that a "reason" to breed them has to be found, as they may not naturally produce as much milk or beef as other breeds. Rather than relying on the kind nature of a few enthusiastic breeders or large conservation programs, actions such as cattle tourism or keeping them as companion animals/pets may be required. In addition, new technology such as breed genetics can help to keep the genetic pool as diverse as possible. Many of these populations have had a genetic bottleneck and avoiding inbreeding is essential.

CATTLE VS. CLIMATE

Probably one of the most heavily debated issues of this century is climate change, and cattle have not escaped criticism. Ironically, both owners and consumers want the same outcome: quality produce at an economically viable price. Feed conversion rates, water intake, use of medicines and antibiotics, and food cost and quantity all affect both the cost of production and the climate. Even the

Below: Of over 3,000 identified modern cattle breeds and hybrids, 196 breeds have become extinct and many are endangered, including the Chillingham wild cattle.

anatomy and physiology of each animal and breed differs, making some more economical and, by default, kinder to the planet. The concerns of farmers, consumers, and those with an interest in our planet are well aligned.

Many modern breeds are producing greater milk yields off less feed energy; likewise, the beef industry is favoring faster-growing breeds with high feed conversion rates. In many parts of the world, people are starting to prefer quality over quantity, which may mean eating less meat but wanting a better cut, and they are often willing to pay more for their choices. Indeed, one area of discussion is whether multipurpose animals are still viable: can cattle really be good at both milk and beef production?

Attempts at making more environmentally viable cows have also been made by changing gut bacteria. For example, using vaccinations to kill methane-producing bacteria while still enabling rumination to occur. It is not just novel research into areas such as vaccines that could help our bovine friends become more eco-friendly. Natural diet changes can also reduce methane production by up to 15 percent; therefore research into feed is essential. Alternatives to fiber could reduce emissions, and even seaweed, probiotics, and other food additives are being tested as possible solutions that could provide significant breakthroughs.

Land loss and deforestation due to beef farming are also of high concern, but some breeds do actually help to conserve land. Making animals more efficient with their feed would also curtail land loss due to grazing.

The range of modern breeds is an ever-expanding topic, with new breeds regularly being recognized throughout the world. The Australian Queenslander started life in 2001 after ranchers teamed up to combat ticks, and it recently became a registered breed. In time, factors such as genetic alteration may well come into play, with breeders picking desired traits and adding to one breed's genome using laboratory methods, rather than relying on breeding alone. With each new breed comes uncertainty: whether it will perform well, or have inherent problems or disease susceptibility; how to best care for it; and whether it will be economically viable.

Above: *Research into food minerals, vitamins, general ingredients, and even differing food sources such as seaweed is vital.*

Left: *Bacteria that live naturally within the digestive system, and those that can infect animals, play essential roles in welfare, production, and economics.*

Competition & Show Cows ✌

Long before the days of the Internet and modern communication methods such as the telephone, email, and cell phone, there were various key ways in which you could share the reputation of your herd—word of mouth, written communications, and competitions and shows. The latter, alongside cattle markets, have been an integral part of life for many valuable reasons. Perhaps the most important reason is gathering experts together in one place, including farmers, retailers, veterinary professionals, butchers, researchers, and feed and technology retailers. They provide an opportunity to share animal health and welfare procedures, breeding practices, housing and feed preferences, and general practice hints and tips. Farming has often been an isolated way of life, so having the opportunity to talk to others in the business is vital. It also offers the opportunity to view other cattle, whether in direct competition or as possible breeding or purchase stock—it is always good for farmers to see the product itself and promote their own.

Competitions and shows also enable the comparison and contrasting of cattle—they're a chance to look at other breeds and understand what works well in the same geographical area. It promotes pride in cattle and encourages good animal welfare; farmers can communicate with the public and show how well cared for and valued are their herds. Competitions are also vital for the economy. They offer a way to market animals, show off good quality, and encourage cooperation and discussions between businesses. A prize-winning bull or cow can enhance the value of a single animal, herd, or even breed overnight. So it comes as no surprise that competitions and shows are a big deal for farming, agriculture, welfare, and the public, with some of the oldest shows even supported by royalty across the globe.

Below: *Shows provide an opportunity to share procedures and practices, view and buy cattle, arrange breeding programs, highlight farming to the public, and for industry workers to socialize.*

COST & TIME

Shows can be local, national, or even international, but this brings a lot of preparation. Travel is not easy. Transporting animals in the appropriate manner, whether by foot, in trucks, or by airplane, rail, or boat, is always trying. Animals mixing from differing herds and regions can increase the spread of disease. Owners must also be away from their farm, ranch, or land, which brings its own worries of their other cattle still needing to be cared for. It usually costs time, effort, and money to attend these events. Even the preparation of the animals attending takes time. It is not unusual for some breeds, such as Highland cattle, to have oils and conditioners added to their coats before a show. This gives the breed a fluffy appearance, so frequently commented upon and admired by the public.

With all these factors in mind, to a certain extent competitions and shows have competitors in their own right. Information and statistics about herd health and productivity are more freely available through electronic methods— images from farm cameras and cell phones—and can be sent anywhere in the world in an instant. Communication methods are much easier in the modern age outside of events, although shows are still popular the world over and may continue to be so for some time to come.

Above: *This Highland cow is being shown by its owner in Britain. Judges may check for body conformation and breed standards in addition to general health and welfare.*

Jersey

WEIGHT
Cows 937 lb (425 kg);
bulls 1,500 lb (700 kg)

HEIGHT
Cows 46 in. (118 cm);
bulls 49 in. (125 cm)

**FEMALE FERTILITY/
CALVING AGE**
First calving at
21–22 months

DISTRIBUTION
Worldwide

OTHER NAMES
Miniature Jersey

MAIN USE
Dairy, some draft

FACT
The milk is famous for its high
levels of butterfat and fat,
making it ideal for cheese,
and is slightly yellow in color.

COUNTRY OF ORIGIN Jersey, Channel Islands, UK

This predominantly dairy breed was first developed in 1700 from cattle from
Normandy, France. The islanders of Jersey and Guernsey valued these animals to
such an extent that they were given as a dowry for marriages between the islands'
residents. After that tradition ceased, the breed became isolated for 200 years until
2008, yet they were exported all over the world. It is said to be one of the fastest
growing breeds in the world today. Some cattle in America are larger, and are
frequently called American Jerseys, while the traditional breed may be called
Miniature Jerseys; however, there is also a separate breed called Miniature Jerseys,
which is much smaller.

Cows calf with little dystocia, but newborns and young animals may need more human
intervention than other breeds. Hypocalcemia (milk fever) rates can be higher in dams
and calves than in other breeds, but mastitis rates are low. The bulls can be aggressive
and cause problems, as this is a breed with horns; cows are calmer but can be nervous
and are known for their tongue rolling. Their coat can be any color but is often fawn.
With more than 2 million individuals, this is the second-most common dairy breed in
the world (after Holstein).

Normande

WEIGHT
Cows 1,500 lb (700 kg);
bulls can reach 2,400 lb
(1,100 kg)

HEIGHT
Cows 57 in. (144 cm);
bulls 61 in. (155 cm)

**FEMALE FERTILITY/
CALVING AGE**
First calving at
24 months

DISTRIBUTION
Now present on all continents

MAIN USE
Principally dairy, but produces
good quality, marbled meat

FACT
Produces 1,742 gallons (6,595
liters) of milk per lactation.

COUNTRY OF ORIGIN France

This gentle breed comes from Normandy, France, but many suggest that its ancestors came from Viking breeds. Other theories suggest that it was bred from French Cauchoise, Augeronne, and Cotentine breeds, which no longer exist. Normande cattle are well known for milk and cheese. High levels of beta and kappa casein in their milk help the curdling process needed to create some cheeses—the famous Camembert is made from their milk. Although Normande cattle are known for the cheeses produced from them, their meat marbles well, which makes it tender.

Recognized in around 1800, during World War II their numbers were reduced greatly in France but recovered to 2–3 million presently living in the country. The breed was exported to South America in 1877, primarily as a beef producer. Colombia still has a large population, and Brazil crossbred it to create the popular Normanzu hybrid. This striking animal usually has a white coat and red spots but can vary in color. They are tough, have hardy feet, birth well, produce good quantities of high-protein milk, reach sexual maturity early, have docile temperaments, and can be used for beef. These cattle are well adapted for colder winters as their coat thickens and curls, while the dark pigment in their eyes helps protect from ultraviolet light in the summer.

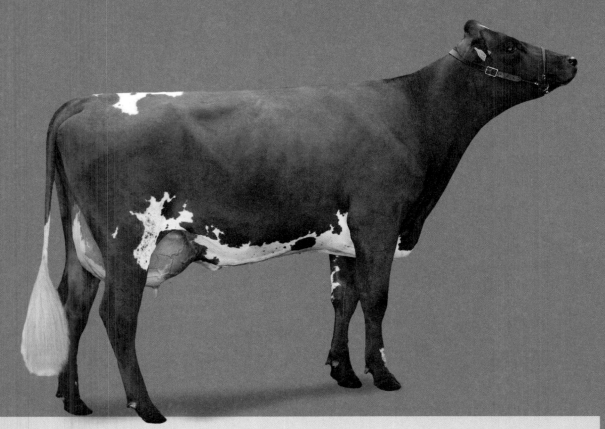

Norwegian Red

WEIGHT
Cows 1,200 lb (550 kg);
bulls 2,900 lb (1,300 kg)

HEIGHT
Cows 54 in. (137 cm);
bulls 57 in. (145 cm)

**FEMALE FERTILITY/
CALVING AGE**
First calving
24–26 months

DISTRIBUTION
Now in over 30 countries

OTHER NAMES
Norsk rødt fe or *Norsk raudt
fe*, NRF

MAIN USE
Dairy, some beef

FACT
Can produce up to 35,000 lb
(16,000 kg) of milk per
lactation.

COUNTRY OF ORIGIN Norway

This breed was founded in 1935 by crossing several Scandinavian breeds including the Norwegian Red-and-White, Red Trondheim, and Red Polled Østland, plus Ayrshires, Friesians, Holsteins, and Swedish Red-and-Whites. While it is a relatively young breed, it became better recognized in the 1970s, and makes up 98 percent of the present day Norwegian herd. Semen and embryo exports are also making the breed popular in the rest of Europe and throughout the world. Despite the name, their color can be red and white or black and white, and although originally they had horns, more of the population are becoming naturally polled through breeding programs.

Norwegians Reds have lower stillbirth and infertility rates than Friesian-Holsteins, and they are often said to be one of the most fertile dairy breeds in the world. Although they produce a lot of milk, mastitis is rare, as is lameness, which is in part due to its hardy hooves. One of this breed's claims to fame is that the makers of the popular Minecraft computer game modeled their cattle on Norwegian Reds. Although predominantly a dairy cow, males can gain around 3 lb (1.4 kg) a day, making it a viable product for the meat industry.

Guernsey

WEIGHT
Cows 1,047 lb (475 kg);
bulls 1,433 lb (650 kg)

HEIGHT
Cows 52 in. (132 cm);
bulls 54 in. (137 cm)

**FEMALE FERTILITY/
CALVING AGE**
First calving at
22–28 months

DISTRIBUTION
Worldwide

OTHER NAMES
Golden Guernsey

MAIN USE
Dairy, historically draft

FACT
Produces a yellow or
gold-colored milk due to
β-carotene, a provitamin
for vitamin A.

COUNTRY OF ORIGIN Guernsey, Channel Islands, UK

Nearly as famous and numerous as Jersey cattle, one of the other Channel Island breeds, the Guernsey now resides throughout the world. One rather famous Guernsey nicknamed the "Sky Queen" became the first cow to fly in an airplane, in 1930, and was also the first cow to be milked on a flight, paving the way for bovines to be transported around the world. This European Blond breed of cattle was registered in 1700, and, similar to the Jersey, other bovines were not imported to Guernsey island for many years, but Guernseys were exported around the world.

The fawn/yellow to reddish-brown coats with white spots are easily recognized, and the Guernsey is renowned for being strong and for converting its feed to produce with a high efficiency. Indeed, 60 percent have the Kappa Casein "B" gene, which gives a firmer curd and increased volume and quality of cheese produced. The milk is also in demand, as it has higher levels of cream (30 percent), protein (12 percent), vitamin D (33 percent), calcium (15 percent), and vitamin A (25 percent) than most other breeds. In addition, 96 percent of this breed have the protein Beta Casein A2, instead of A1, which, some say, makes the produce healthier. Although calves are born large, they have few labor-related complications.

Gyr

WEIGHT
Cows 849 lb (385 kg);
bulls 1,202 lb (545 kg)

HEIGHT
Cows 51 in. (130 cm);
bulls 55 in. (140 cm)

**FEMALE FERTILITY/
CALVING AGE**
First calving at
3–4 years

DISTRIBUTION
India, South America,
North America

OTHER NAMES
Gir, Bhodali, Desan, Gujarati,
Kathiawari, Sorthi, Surti

MAIN USE
Dairy, some beef

FACT
Can produce milk containing
up to 4.5% fat.

COUNTRY OF ORIGIN Gujarat, India

The Gyr are highly distinctive as they have domed, rounded foreheads (ultraconvex), which makes them unique among cattle of the world. In addition, they have an extra layer of muscle under their skin, enabling them to shake their skin and help prevent parasites. These cattle range from white, yellow, and/or red, but not black, and are generally mottled with the undercoat being red. They also produce oils which help protect them from insect bites and have impressive spiraling, sweeping horns and long ears. Local people known as Maldhari traditionally kept this breed in India, but sadly buffalo milk popularity has decreased numbers of Gyr.

This zebu breed originated in India, where it is still a principle zebu breed, and is now in countries such as Brazil. Genetic studies have indicated that 50 percent of the population presently in Brazil came from just 28 original ancestors, which in turn has resulted in careful breeding, to ensure large genetic variation. Gyr are often crossed with Friesians, in order to increase milk yield and to produce Girolando, which produce 80 percent of Brazil's milk. They are also used to develop the Brahman breed in North America. Most known for its milk, this breed performs differently depending on its environment, genetics, and farming style.

Belarus Red

WEIGHT
Cows 1,102 lb (500 kg);
bulls 1,874 lb (850 kg)

HEIGHT
Cows 51 in. (129 cm);
bulls 57 in. (145 cm)

**FEMALE FERTILITY/
CALVING AGE**
First calving at
26 months

DISTRIBUTION
Belarus

OTHER NAMES
Byelorussian Red

MAIN USE
Dairy (some beef)

FACT
High milk fat content,
averaging 4.32% fat,
increasing further in the
top-producing herds.

COUNTRY OF ORIGIN Belarus

This Belarusian dairy cattle breed has been developed over the last century by breeding programs starting with the Angeln Red and German Red. The Polish Red and Danish Red were introduced in the 1920s and 30s, followed by the Estonian Red and Latvian Brown in the 1950s, and more recently with the Danish Red again. As the name suggests, these cattle are common in Belarus, especially around Grodno and Minsk, and are red or rust-red in color. The medium-sized breed has striking horns, a fairly long face, and moderate levels of musculature, which enables them to be used for beef.

The Belarus Red has adapted well to the local climate, has a lengthy lifespan, and copes well with a varied diet. However, milk production rates are affected by lower-quality food. Overall, milk yields have increased over the decades as food quality and breeding have improved. In the 1980s, average milk yield was 5,637 lb (2,557 kg) rising to 6,731 lb (3,053 kg) in the top-producing farms. More recently, the averages have increased to 9,952 lb (4,514 kg) of milk, and 13,351 lb (6,056 kg) for the best-producing cow.

Girolando

WEIGHT
Cows 895 lb (406 kg);
bulls 1,430 lb (650 kg)

HEIGHT
Cows 50 in. (127 cm);
bulls 55 in. (140 cm)

**FEMALE FERTILITY/
CALVING AGE**
First calving at
30+ months

DISTRIBUTION
Brazil

OTHER NAMES
Gyr Holstein Cross

MAIN USE
Dairy, some beef

FACT
Can produce 44,000 lb
(20,000 kg) of milk across
a 15-year period.

COUNTRY OF ORIGIN Brazil

This breed is a *Bos taurus* crossed with *Bos indicus* in the form of Holstein crossed with Gyr (also called Gir). The Gyr is native to India and is well adapted for tropical climates, while the Holstein is well known for its milk production capacity. These attributes make a good combination for dairy herds in Brazil, where the Girolando was established. The genetic makeup is around ⅜ Gyr and ⅝ Holstein, and usually the cattle produced are black and white, but can be brown, similar to the Holstein but with large ears like the Gyr. They have a docile nature, making them easy to handle.

Despite being a young breed, only developed in the 1940s and officially registered in 1989, it is already popular throughout Brazil. Girolando are excellent dairy cattle, with up to 100 percent fertility levels in some herds and able to produce around 80 percent of Brazil's milk. Peak milk production is reached at around 10 years old. Although this breed is mostly used for dairy, with a weight gain of up to 2 lb (1 kg) per day it can be used for beef. Female calves have a much higher weight gain than males initially, similar to some Friesians. As Brazil produces so much milk and has such a high number of cattle, it has looked at the entire DNA sequence of this breed alongside others to gain further insight into its traits.

Sahiwal

WEIGHT
Cows 937 lb (425 kg);
bulls 1,300 (600 kg)

HEIGHT
Cows 47 in. (120 cm);
bulls 54 in. (136 cm)

**FEMALE FERTILITY/
CALVING AGE**
First calving at 39 months

DISTRIBUTION
Pakistan; India, Australia,
Bangladesh, Africa, Caribbean

OTHER NAMES
Mint kumre, Lambi Bar, Lola,
Montgomery, Multani, Teli

MAIN USE
Dairy; beef and draft too

FACT
Usually feed their young,
therefore 5,000 lb (2,270 kg)
of milk produced per lactation
is reasonable.

COUNTRY OF ORIGIN Pakistan

This breed is named after the Sahiwal district in Pakistan where it originated and still resides. Presently, it is protected by the government, as it is considered endangered and, as an indigenous breed, is valued in the region. Carefully managed by herdsmen called *charwahas*, incentives are given by the government to keep Sahiwal. The Research Center for Conservation of Sahiwal Cattle has established a breeding program. This red to dun-colored animal often has darker colors around its well-developed thoracic hump and neck, and has long dropping ears.

This zebu breed is said to produce the most milk of all zebus. Although known for its dairy capacity, it was originally a dual-purpose animal producing fast-growing young which were good for the beef industry. It is also strong, making it useful for pulling carts, although it is known for being lethargic, so may not get you anywhere fast. Cattle exported to Australia were used for meat, and were often crossed with other beef breeds to enhance their qualities, especially for heat tolerance and resistance to ticks. The breed is highly valued in Kenya among the Maasai people. Surviving very arid lands, it produces around 10 percent of the country's milk.

Brown Swiss

WEIGHT
Cows 1,320 lb (600 kg);
bulls 1,980 lb (900 kg)

HEIGHT
Cows 57 in. (145 cm);
bulls 59 in. (150 cm)

**FEMALE FERTILITY/
CALVING AGE**
Fertile at 26 months

DISTRIBUTION
Worldwide

OTHER NAMES
American Brown Swiss;
Braunvieh

MAIN USE
Dairy, a little beef

FACT
Milk contains smaller fat
globules and longer-chain
fatty acids.

COUNTRY OF ORIGIN USA

In spite of the name, this breed was developed in the USA, but from the Braunvieh breed from Switzerland. Although the Braunvieh is used for beef, dairy, and draft, the Brown Swiss has been bred selectively for dairy traits. With milk yields over 22,000 lb (10,000 kg) per annum, peak yield is often reached on or after fifth lactation, and yet mastitis levels are low. The Brown Swiss is now very different from its ancestors, which may have a history spanning over 4,000 years, particularly as it produces so much milk. As the milk contains 3.5 percent protein and 4 percent butterfat, it is known for producing high-quality cheeses. It also produces high levels of A2 beta-casein protein rather than A1. A2 milk has become popular in many countries, with many non-bovine milks containing predominantly A2. As inbreeding for milk yield became more intensive, certain genetic defects became more prevalent in this population, including weaver disease and spinal muscular atrophy. Apart from these concerns, the Brown Swiss remains popular for crossbreeding.

The animals can have short horns or be naturally polled. Their coat color is usually gray, brown, tan, and/or white, and their ears are fairly unusual as they are fluffy, long, and large. Originating from draft cattle ensures a robust animal.

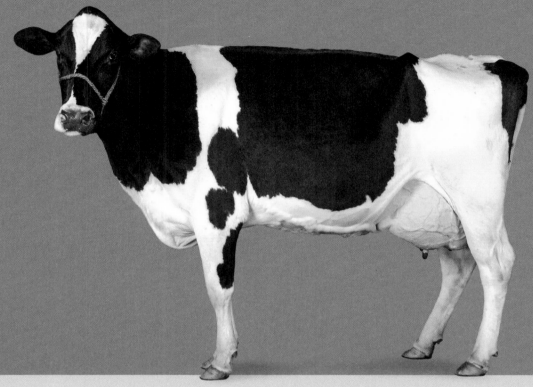

Holstein-Friesian

WEIGHT
Friesian cows 1,280 lb (580 kg), Holstein cows 1,600 lb (725 kg); bulls 2,000 lb (907 kg)

HEIGHT
Cows 61 in. (155 cm); bulls 72 in. (182 cm)

FEMALE FERTILITY/ CALVING AGE
First calving at 22–24 months

DISTRIBUTION
Worldwide

OTHER NAMES
Holstein, Friesian

MAIN USE
Dairy, but also used for beef

FACT
Milk production from these cattle has doubled over the last four decades.

When people think of cows, the one that often comes to mind is the Holstein-Friesian, famous for being the highest producer of milk. This iconic animal can be black and white piebald, sometimes red and white, and, rarely, a "blue" breed (white in black hair gives this illusion). They have horns or can be naturally polled. In the modern breed, the Holsteins refer to North American cattle, and Friesians to those from European bloodlines. Some suggest that 87.5 percent of Holstein blood and more than 12.5 percent of Friesian blood makes the animal Friesian, the reverse is true for Holstein. Dating back to ancestral breeds from around 100 BCE, this animal has undergone careful breeding programs.

This breed has produced a number of famous, even celebrity, cattle, including Pauline Wayne, US President Taft's cow, and record-holding Bur-Wall Buckeye Gigi EX-94 3E, who produced 74,650 lb (33,860 kg) of milk in one year. A heifer named Daisy was the first Holstein created as a clone in 1999 (the fifth ever cattle clone born alive), quickly followed by clones Amy, Betty, and Cathy. A well-known clone of CIAQR sire Hanoverhill Starbuck was born in 2000 and called Starbuck II. One bull named Toystory has sired a remarkable 500,000 plus offspring from over 2.4 million semen straws.

COUNTRY OF ORIGIN The Netherlands and north Germany

Icelandic

WEIGHT
Cows 950 lb (430 kg);
bulls 1,300 lb (600 kg)

HEIGHT
Cows 50 in. (127 cm);
bulls 53 in. (135 cm)

**FEMALE FERTILITY/
CALVING AGE**
First calving at
20–24 months

DISTRIBUTION
Iceland

OTHER NAMES
Íslenska kýrin

MAIN USE
Dairy, some beef

FACT
Cows produce 13,200–
24,000 lb (6,000–11,000 kg)
of milk per annum.

COUNTRY OF ORIGIN Iceland

These native animals originate from the ancestors brought over by the Vikings when they settled a thousand years ago. In order to remain disease free, importation of cattle is banned. Around 95 percent of the population are naturally polled, and the animals are kept in sheds for 8 months of the year due to the cold weather. With the environment being relatively hostile, with sparse vegetation and cold winter months, it is easy to see why this country protects its valuable cattle.

Despite producing decent yields of milk, some have suggested importing or even replacing this native population with higher-production animals, but these cattle are a big part of Icelandic history. Their milk makes cheese and the popular *skyr*, a fresh, sour-milk cheese eaten worldwide as a yogurt, similar in consistency to Greek yogurt. Popular in Iceland for centuries, this healthy food likely takes its name from the English word "shear" (to cut), as the liquid whey and thick skyr are split during the processing technique. Many farms still make *skyr* independently using traditional methods. The Icelandic breed, numbering around 75,000, resides only in Iceland. The famous Icelandic Phallological Museum collection even started with specimens from bulls.

Illawarra

WEIGHT
Medium to large size

HEIGHT
Medium to large size

**FEMALE FERTILITY/
CALVING AGE**
First calving at 24 months

DISTRIBUTION
Australia, North America,
Indonesia, Japan, Korea,
Middle East, New

Zealand, Pacific Islands,
Pakistan, UK

OTHER NAMES
Illwarra Shorthorn, Australian
Illawarra Shorthorn

MAIN USE
Dairy, some beef

FACT
Able to produce over
10 gallons (40 liters) of
milk a day.

COUNTRY OF ORIGIN Australia

This Australian breed was developed from Shorthorns and various other breeds, and takes its name from the region in which it originated, Illawarra, in New South Wales. This area of land was originally cleared for farming by early settlers and convicts, producing good ranching land. In 1899 some Illawarra were registered in the *Milking Shorthorn Herdbook*, but the official breed herdbook was established in 1910, after a few decades of breeding these cattle.

Mostly red, with some white or roan animals, Illawarra have horns in both sexes, but are generally docile. They are able to withstand the hot summer conditions, reaching over 104°F (40°C) down to sub-zero, and are efficient at feeding, requiring few supplements to their diet. Their dark hoof and skin pigmentation also help to protect from the sun. Their pelvis shape is well suited to birthing, therefore it is rare that assistance is needed. Producing a calf every year into her teens, a cow can produce in excess of 16,000 gallons (60,000 liters) of milk in her lifetime. This now popular breed has its own society and shows.

Belmont Red

WEIGHT
Cows 990 lb (450 kg);
bulls 1,400 lb (650 kg)

HEIGHT
Cows 47 in. (120 cm);
bulls 48 in. (121 cm)

**FEMALE FERTILITY/
CALVING AGE**
First calving at
32–34 months

DISTRIBUTION
Australia

MAIN USE
Beef

FACT
Consistently wins against
other breeds in the Aus-Meat
Comparison Trials started
in 1988.

During the 1950s, Australia was concerned about the fertility of *Bos indicus* cattle. The Commonwealth Scientific and Industrial Research Organization (CSIRO) responded by crossbreeding Africander (African Sanga) and Hereford-Shorthorn (*Bos taurus*) to create the Belmont Red. This cross is named after the research station at which it was developed, and the breed became available to farmers in the late 1960s. As the name suggests, the coat color is generally red but can be white. In either case, the coat is smooth and sleek with short hair, and so the breed can sweat more, making it more appropriate for and tolerant to the Australian heat.

Not only does this breed have good fertility rates, but it is also highly resistant to ticks (which represents an economic saving as they have to be dipped less for tick control), has quality meat and feed efficiency, grows well, and has a good temperament. The percentage of live sperm in semen is 79 percent, much higher than the minimum acceptable level for cattle of 60 percent. All of these assets make Belmont Reds ideal for Australian farming systems. The breed consistently scores well in Australia for its meat and is of a better standard than any other breed or crossbreed in the country.

COUNTRY OF ORIGIN Australia

Gelbvieh

WEIGHT
Cows 1,500 lb (700 kg);
bulls 2,600 lb (1,200 kg)

HEIGHT
Cows 55 in. (140 cm);
bulls 60 in. (152 cm)

**FEMALE FERTILITY/
CALVING AGE**
First calving at 22 months

DISTRIBUTION
Europe, Canada, USA,
Australia, Africa

OTHER NAMES
German Yellow, Yellow Cattle;

MAIN USE
Beef (originally beef, dairy,
and draft)

FACT
Of all the beef breeds in the
world, the Gelbvieh has the
largest ribeye muscle area per
220 lb (100 kg).

COUNTRY OF ORIGIN Germany

Gelbvieh (pronounced *Gel-fee*) were originally bred in Bavaria, southern Germany. The name translated from German means "yellow cattle." This breed was developed in the eighteenth to nineteenth centuries from several local breeds for use in the beef, milk, and draft industries. Eventually, through crossbreeding, the many strains were amalgamated into one breed. Gelbvieh was officially recognized as a breed in the 1920s and is best known today for beef production. Originally reddish golden brown to russet in color with horns, many cattle have been selectively bred to be polled and some have black coats. They grow quickly and efficiently, easily converting their food into growth rates, are sexually mature at a young age, and the offspring are small, enabling easier calving for the dams, who produce good milk.

Modern day breeders often cross Gelbvieh with other cattle, but to remain "pure" they must remain 88 percent Gelbvieh. This is especially common in the USA following extensive research programs in Nebraska. Producing good quality lean beef, and especially tender meat as yearlings, the bulls are often crossed with Angus females to produce prize-winning beef. Reports from South Africa and Australia confirm that they are heat tolerant and tick resistant.

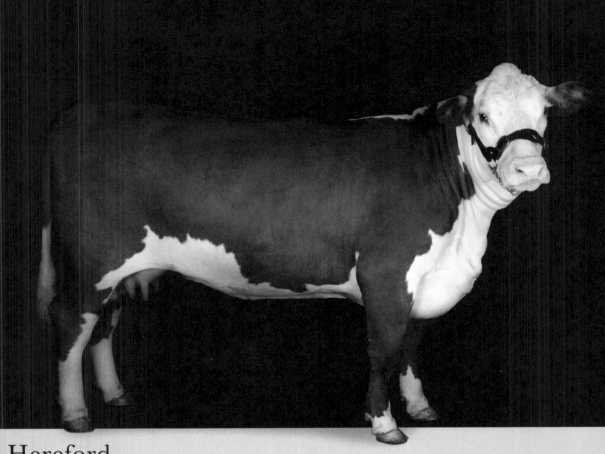

Hereford

WEIGHT
Cows 1,800 lb (800 kg);
bulls 2,600 lb (1,200 kg)

HEIGHT
Cows 55 in. (140 cm);
bulls 60 in. (152 cm)

**FEMALE FERTILITY/
CALVING AGE**
First calving at
24–36 months

DISTRIBUTION
Over 120 countries worldwide

OTHER NAMES
Whiteface

MAIN USE
Beef

FACT
This natural grazer numbers
10 million (5 million pure
breeds), producing high
quality marbled beef.

COUNTRY OF ORIGIN England, UK

With over 5 million Hereford purebreds throughout the world, this bovine is extremely well known and popular. First bred in England, it took its name from the county it derived from, Herefordshire. During the eighteenth century, this breed was bred with Shorthorns and other local breeds including Welsh border cattle. By the nineteenth century, the characteristic white face and typical red to red-yellow coloring had been established. Exported in the 1800s to the USA, Canada, and South America, it now thrives from the cold, harsh steppes of Russia and the colder Scandinavian climes through to the drier, hot environments in Australia, Africa, and South America. Eye cancer is more prevalent for those animals in countries with more sunlight, as is udder sunburn, and dwarfism is a genetic trait observed in this breed throughout the world.

With early sexual maturity, short gestation times, a docile nature, high libido in males, and protective maternal natures, these cattle are easy to breed. The Black Hereford was derived by crossing the Hereford with Black Angus and is recognized as a breed in its own right. Miniature Herefords are about a third of the size of the traditional Hereford; they are not a dwarf breed and were deliberately bred for small stature.

Belgian Blue

WEIGHT
Cows 1,900 lb (875 kg);
bulls 2,590 lb (1,175 kg),
can reach over 2,900 lb
(1,300 kg)

HEIGHT
Cows 55 in. (140 cm);
bulls 58 in. (148 cm)

**FEMALE FERTILITY/
CALVING AGE**
First calving at 32 months

DISTRIBUTION
Europe, Brazil, USA, Canada,
New Zealand

OTHER NAMES
Belgian Blue-White, Belgian
White and Blue Pied, Belgian
White Blue, Blue

MAIN USE
Beef

FACT
Carcass yield up to 80%.

COUNTRY OF ORIGIN Belgium

The Belgian Blue originates from Belgium and was developed from around 1850–1890. Similar to the Italian breed called Piedmontese, it has enhanced levels of muscle, so called "muscle doubling." Therefore, they have low levels of fat and are lean. A genetic mutation in the myostatin gene is responsible for building muscle levels and reducing body fat retention. In order to fully benefit from this mutation, older animals need higher protein levels in their diets in comparison to other breeds. Unfortunately, as the calves have such a large birth weight, cesarean sections are the norm when crossing pure breeds, in order to prevent dystocia/problematic births.

This breed usually needs longer growth periods before it can go to market for meat, and it matures at a relatively older age in comparison to some other breeds. Due to the increased costs of feed, farm management, and veterinary care, these animals can be more difficult to justify on an economic basis in comparison to other breeds. Initially used as both beef and dairy breeds, it became bred for beef specifically in the 1950s, after breed models were established based on three sires (Gedeon and two of his grandsons, Ganache and Vaiseur). By the 1970s this popular breed had spread throughout the world.

Limousin

WEIGHT
Cows 1,650 lb (750 kg);
bulls 2,540 lb (1,150 kg)

HEIGHT
Cows 55 in. (140 cm);
bulls 58 in. (148 cm)

**FEMALE FERTILITY/
CALVING AGE**
First calving at
24–32 months

DISTRIBUTION
Over 70 countries

OTHER NAMES
Limousine

MAIN USE
Beef, some sire dairy cattle

FACT
Despite early records of poor
beef quality, this breed was
starting to win awards for
Best European Breed by 1857.

COUNTRY OF ORIGIN France

Created in a region in France, this breed of cattle was named after their original home provinces (Limousin and Marche) in 1886, but records indicate the cattle were probably present in the region in 1770. DNA studies have shown links to many other breeds, mostly of European ancestry. In the early nineteenth century the records show that this now mighty beast was a mere 715 lb (325 kg) but by 1862 weights of 1,320 (600 kg) were recorded. After a rapid decline, the 1960s saw an attempt to revive the breed and today it is the second-most common beef breed in France, outnumbered only by the Charolais.

The introduction of polled animals has been controversial, with breeders having difficulties in getting them recognized for registration. Myostatin gene variations means that this breed has excellent muscle mass, not quite as much as the Belgian Blue and Piedmontese but more than other prime beef cattle. Not every animal has this mutation, but around 90–96 percent of full-bloods have the mutation in both alleles (called homozygous), giving optimal muscle mass and passing the trait on to offspring. Up to 80 percent of the carcass yields meat. The usual coat color is light wheat to darker golden-red, but other colors such as black are recognized.

Parthenais

WEIGHT
Cows 1,900 lb (875 kg);
bulls 2,500 (1,150 kg)

HEIGHT
Cows 57 in. (145 cm);
bulls 61 in. (155 cm)

**FEMALE FERTILITY/
CALVING AGE**
First breeding at
26–28 months

DISTRIBUTION
UK, Ireland, Belgium,
Netherlands, Canada, USA

OTHER NAMES
Gâtinaises, Boeufs de Gâtine

MAIN USE
Beef

FACT
Meat from this breed sells at
a premium price, 25% more
than other European breeds.

One of the world's largest cattle, the Parthenais breed was originally used for beef, draft, and dairy, even producing a type of butter called "Charente-Poitou." It originated in central France, is named after the local town Parthenay, and formally established its own herdbook in 1893. As a multipurpose animal, other breeds such as Friesian, Normande, and Charolais were outperforming Parthenais in the dairy and beef industries. As the need for draft cattle in agriculture reduced, the breed was refined over the last seven decades for the beef industry, in order to remain competitive and economically viable.

Its coat is red, golden brown to dark brown, with some black around the head and neck region and on the tail, with hard black feet and horns in both sexes. Although not quite as large as the Belgian Blue, Parthenais show double muscling but benefit from easier calving, even if they mature later than other heifers. Double muscling in these animals results in much larger muscles, leading to more muscle in both calves and adult cattle. Despite being large, these animals are mobile, fertile, hardy, and fare well in differing climates. With only 8.8 percent fat on the carcass, the tender meat is seen as a healthy beef and has lower cholesterol than chicken.

COUNTRY OF ORIGIN France

Beefmaster

WEIGHT
Cows 1,100 lb (500 kg);
bulls 2,600 lb (1,200 kg)

HEIGHT
Cows 51 in. (129 cm);
bulls 53 in. (135 cm)

**FEMALE FERTILITY/
CALVING AGE**
First calving by 24 months

DISTRIBUTION
USA, Brazil, South Africa

OTHER NAMES
"The Profit Breed"

MAIN USE
Beef, some dairy

FACT
Whether the temperatures are
very high or very low, this
animal will produce good
quality products.

COUNTRY OF ORIGIN USA

This breed was developed in Texas. By establishing a breeding program running from 1908 to the 1930s, the family-owned ranch developed the cattle by crossbreeding Hereford and Shorthorn cows with Brahman bulls. By the 1950s the Beefmaster was officially recognized as a breed, and the family had moved the herd to Colorado, where the breed was able to withstand colder climates. The breed was developed to withstand the high temperatures of Texas, and to this day they have a high heat tolerance—they do not even take to the shade in the hottest parts of the day; they continue grazing. Calves generally weigh in at around 77 lb (35 kg); the females calve easily and yearly, and produce good amounts of high quality milk. Males breed readily and are intelligent and docile, making them easy to handle and mate.

Ed and Tom Lasater, who founded the breed, decided that looks were not as important as production traits. The exact pedigree of the breed remains unknown, but the Lasater family concentrated on the so-called "Six Essentials." By concentrating on Disposition, Fertility, Weight, Conformation, Hardiness, and Milk Production, they founded a breed that is now the fourth most popular in the USA. Beefmaster tend to be red in color with mottled white spots, and can be horned or naturally polled.

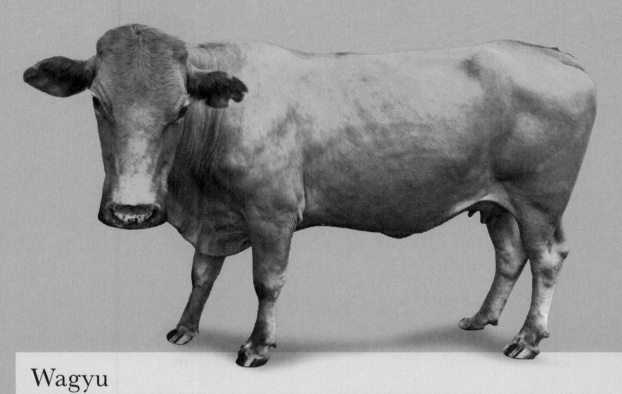

Wagyu

WEIGHT
Cows 1,129 lb (512 kg);
bulls 1,784 lb (809 kg)

HEIGHT
Cows 51 in. (129 cm);
bulls 57 in. (145 cm)

**FEMALE FERTILITY/
CALVING AGE**
First calving at
22–24 months

DISTRIBUTION
Australia, USA, Canada, UK

OTHER NAMES
Washu

MAIN USE
Beef

FACT
Produces some of the most
expensive beef in the world.

COUNTRY OF ORIGIN Japan

Literally translated as "Japanese cow," Wagyu is the generic name for any of the four Japanese beef breeds: the Japanese Black, Japanese Brown, Japanese Shorthorn, and Japanese Polled. The Black is the most popular breed, making up 90 percent of the beef cattle in Japan; Brown 9 percent; Shorthorn 1 percent; and Polled are rare. Both sexes have horns except for the Polled breed, in which horns are absent in both sexes. Wagyu cattle descended from native Japanese breeds crossed with European breeds in the twentieth century. Wagyu meat has very fine marbling, making it renowned throughout the world.

The Japanese Black has several strains, including the *Tajima* strain, the beef of which can be sold as "Kobe" beef if reared in the Hyōgo Prefecture. Kobe is an expensive delicacy, with restaurant customers paying $170–500 per pound, and has only been exported since 2012. General high-grade Wagyu beef can come in at around $300 per pound; Wagyu cows sell for around $30,000, more than ten times the amount of the popular Angus breeds.

Angus

WEIGHT
Cows 1,200 lb (550 kg);
bulls 1,900 lb (850 kg)

HEIGHT
Cows 46 in. (118 cm);
bulls 53 in. (135 cm)

**FEMALE FERTILITY/
CALVING AGE**
First calving at
22–24 months

DISTRIBUTION
Australasia, Europe, southern
Africa, North & South America

OTHER NAMES
Aberdeen Angus, Black
Angus, Doddies, Hummlies,
Red Angus

MAIN USE
Beef

FACT
Good beef marbling qualities.

Originally developed in Aberdeen, Scotland, this popular beef breed is sometimes classified as Black Angus (the original color) or the more recent Red Angus, depending on the coat color. Neither cows nor bulls have horns, but this hardy, medium-sized breed has been around since the sixteenth century. The Black Angus has become the most common beef breed in the USA and the most populous native breed in the UK. One in four Australian cattle are registered as Angus, and they are a dominant breed in countries such as Argentina, Canada, and New Zealand. Part of the reason it has become so popular is its link with many large fast-food companies: recognizing good quality beef, coupled with excellent availability and affordability, Angus meat has been used by many international restaurants and franchises.

DNA tests are available for four recessive genetic disorders which are common in the breed, mainly affecting the bones and joints. The cows make good mothers, and gestation is a couple of weeks shorter than most European breeds. In addition, they grow well, so are ready for slaughter relatively quickly. Bulls can have a reputation for aggression; appropriate handling is especially important in males. The popular Brangus breed originating in the USA was developed by crossing Angus with Brahman.

COUNTRY OF ORIGIN Scotland

Beefalo

WEIGHT
Cows 1,500 lb (680 kg);
bulls 2,000 lb (907 kg)

HEIGHT
Cows 55 in. (140 cm);
bulls 56 in. (142 cm)

**FEMALE FERTILITY/
CALVING AGE**
Fertile at 24–36 months

DISTRIBUTION
USA, Canada, Australia

OTHER NAMES
Cattalo, Canadian hybrid

MAIN USE
Beef

FACT
Meat has 66% fewer
calories, 79% less fat, and
cholesterol levels are a third
lower than beef cattle,
chicken, and even cod, while
retaining high protein levels.

COUNTRY OF ORIGIN USA

This fascinating breed is a cross between *Bos taurus* (usually male) and American bison. The offspring can look more like cattle than bison, but officially should contain a maximum of 37.5 percent of bison genetic information. Deliberately bred since the nineteenth century, natural offspring were noted as far back as 1749. These early animals were often called cattalo, although this can also refer to animals with more than 37.5 percent bison. Various breeding programs have been tried over the years, combining males and females from the differing animals, and looking at the resulting number of offspring, their traits, and fertility. The main aim of this breed was to achieve a hardy animal which produced a good milk yield and beef quality. They calf independently and are more docile than bison, yet grow quickly and can tolerate cold weather due to a thick coat.

Over the years there has been some concern over bison conservation. There are only four herds today which do not have some levels of cattle DNA. This breed is being seen as an ideal animal for consumers who prefer to see free-range, antibiotic-free, and steroid-free produce. The Beefalo can be a good investment for the rancher as they can cost 40 percent less than other cattle to produce.

Texas Longhorn

WEIGHT
Cows 990 lb (450 kg);
bulls 1,800 lb (800 kg)

HEIGHT
Cows 48 in. (122 cm);
bulls 60 in. (152 cm)

**FEMALE FERTILITY/
CALVING AGE**
Breeding from 12–13
months, calving at just
over 24 months

DISTRIBUTION
Now imported to other
countries

OTHER NAMES
American Longhorn

MAIN USE
Beef, steer riding, shows,
draft work

FACT
The meat is lean with low
cholesterol levels.

At around 6 ft (1.8 m) from tip to tip, and a record-breaking bull with a total horn length of 10.8 ft (3.29 m), both males and females have horns. This breed has an impressive history, descending from the original cattle brought to the New World by Columbus in 1493. The Spanish later imported more, which roamed wild, surviving for hundreds of years in the area that later became Texas. The Texas Longhorn became a hardy, self-sufficient animal able to live off sparse lands, eating weeds, surviving droughts, and roaming for miles. After being bred with feral Mexican cattle the breed was tamed and lived in ranches.

In the nineteenth century, Texas Longhorns would be driven up to 1,500 miles (2,500 km) to market. Most drives contained around 2,000 cattle, but the largest recorded was 15,000, in 1869. Later, the Texas Longhorn became less popular as it produced a lean meat when tallow (made from fat) was in demand. It is hard to believe that this highly prized breed, which can now be sold for close to $400,000 (more usually $40,000 for a prize bull), nearly became extinct. The U.S. Forest Service mounted a rescue operation, and presently the breed's conservation status is classified as "critical."

COUNTRY OF ORIGIN USA, Canada

Highland

WEIGHT
Cows 1,000 lb (470 kg);
bulls 1,500 lb (700 kg)

HEIGHT
Cows 39 in. (100 cm);
bulls 46 in. (117 cm)

**FEMALE FERTILITY/
CALVING AGE**
First calving at
24–36 months

DISTRIBUTION
Worldwide

OTHER NAMES
Long-haired, North Highland,
Scottish, West Highland

MAIN USE
Beef, some dairy

FACT
Highland cattle's milk has
a high butterfat content.

COUNTRY OF ORIGIN Scotland

When the meat is good enough for the British royal family to keep a herd of its own, you know you have an excellent beef breed. In 1954, Queen Elizabeth II started keeping a "fold" in her Scottish home, Balmoral Castle. Highland cattle are said to live in folds, rather than herds, as they are often kept in open stone shelters called folds. Descending from the Hamitic Longhorn, two highland breeds called West Highland and Mainland originally existed. By around 600 CE, they had been crossbred and the remaining breed is simply termed Highland cattle.

This hardy breed has distinctive features. Its long horns and double wooly coat covered with waterproofing oils, the longest hair seen in any breed, make it a popular breed throughout the world. Best known for its coat colors with red hues, Highland cattle range from reds and browns to black in color. These differences are due to differing gene sequences in at least three genes (to date) called MC1R, PMEL, and SILV. Coat thickness and other adaptations such as food scavenging skills have made them very capable of living in cold climates. They have been compared to Arctic-dwelling animals such as reindeer for their ability to tolerate low temperatures. Their gentle temperament makes them relatively easy to handle.

Galloway & Belted Galloway

WEIGHT
Cows 1,200 lb (550 kg);
bulls 1,800 lb (800 kg)

HEIGHT
Cows 47 in. (120 cm);
bulls 53 in. (135 cm)

**FEMALE FERTILITY/
CALVING AGE**
First calving at
29–36 months

DISTRIBUTION
Australia, Brazil, Canada,
France, Germany, USA,
Netherlands, New Zealand,

OTHER NAMES
Beltie, Sheeted Galloway,
Oreo, White-middled

MAIN USE
Beef (some cheese)

FACT
Produce high quality beef

COUNTRY OF ORIGIN Scotland

Galloways were first developed in the seventeenth century in Scotland and number around 10,000 throughout the world today. Galloway cattle have no horns despite originating from horned cattle. Cattle drovers preferred handling the polled animals, so through selective breeding it became a feature of the breed. They also have a double-layered coat with guard hairs which help with waterproofing the animal and ensuring insulation; the coat sheds in warmer weather. The Galloway is usually a dun, black, or red color as a result of mutations PMEL gene (dun) or alleles of the MC1R gene for black or red.

The Belted Galloway was officially established in 1921, derived directly from the Galloway possibly crossed with the Dutch Lakenvelder. It has a distinctive white band of hair around its middle. Exported worldwide, there are around 18,390 in the USA, five times more than in Scotland itself. The numbers in Scotland and the UK decreased dangerously during the 2001 foot and mouth disease outbreak, but have since recovered. This breed has regular shows and even an annual congress. The Galloways have enjoyed a revival as demand for quality meat has increased—their ability to produce high quality meat despite poorer grazing areas is in demand.

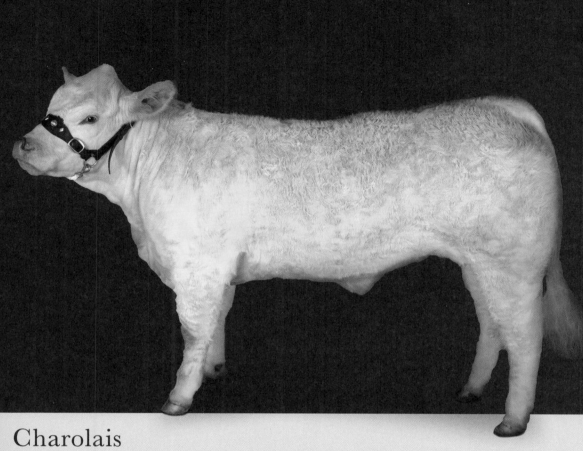

Charolais

WEIGHT
Cows 2,200 lb (1,000 kg);
bulls 3,000 lb (1,400 kg)

HEIGHT
Cows 53 in. (135 cm);
bulls 67 in. (145 cm)

**FEMALE FERTILITY/
CALVING AGE**
First calving at 24 months

DISTRIBUTION
Worldwide distribution

started in the 1930s and
increased from the 1950s

OTHER NAMES
Charolaise

MAIN USE
Beef, were used for milk and
draft work

FACT
Renowned for its meat and fat
colors and low fat percentage.

COUNTRY OF ORIGIN France

This white- or cream-coated breed, with horns, pale hooves, and a pink muzzle has a distinctive phenotype. More recently, black- and red-coated varieties have been created. Developed in France around 878 CE, it is still the most common French beef breed, with over 4 million in the country. With an estimated 750,000 across 68 other countries worldwide, these well-muscled animals are very popular. With bull weight rates hitting up to 4,400 lb (2,000 kg) it is certainly a large beast—added to that their horns and more aggressive temperament, care must be taken when handling this breed or housing males together. Maternal instinct is strong and can result in aggression when threatened. This large breed should not be crossed with smaller breeds, as smaller dams can be at risk of dystocia, problems in labor, and uterine prolapse.

The Charolais breed has adapted well to colder weather and has a shorter coat in summer, making it fairly heat tolerant, but it is not necessarily well adapted for lots of sun. Its light coloring makes it more likely to get sunburn, and its eyes are more prone to cancer than other breeds. Charolais leukodystrophy/Inherited Progressive Ataxia is an inherited trait seen in this breed, as is cleft palate and a few other disorders; but otherwise they are a hardy breed.

Devon

WEIGHT
Cows 1,250 lb (570 kg);
bulls 2,500 lb (1,100 kg)

HEIGHT
Cows 51 in. (130 cm);
bulls up to 69 in. (175 cm)

**FEMALE FERTILITY/
CALVING AGE**
First calving at
24–36 months

DISTRIBUTION
Present on five continents

OTHER NAMES
North Devon: Devon Ruby,
Ruby Red; South Devon:
"Gentle Giant"

MAIN USE
Beef, some milk

FACT
Originally a dual-purpose
animal.

COUNTRY OF ORIGIN England

Devon breeds are generally split into North Devon and the more recent South Devon. The ancient North Devon cattle have red or tawny coats, whereas the more recent South Devon breed has a yellow/brown coat. The original North Devon cattle were traditional red bovines of southern England, crossed with Hereford, Sussex, Lincoln Red, and Red Poll. They are a strong and sturdy breed with a high resistance to parasites and disease. The American Milking Devon, developed from the North Devon breed dating back to the 1600s, nearly became extinct in the 1970s and is still endangered, but preservation efforts are in force.

The South Devon breed is one of the largest breeds; the heaviest recorded animal weighed in at 4,400 lb (2,000 kg). The South Devon cattle are said to have descended from a red Normandy breed following a French invasion. Using larger males for breeding has increased the incidence of difficult births, but they are hardy and birth easily. They also have good mothering skills and provide excellent milk for their calves. Their long winter coat and sleek summer coat enable them to adapt to differing climates, and they produce high quality beef and milk.

Droughtmaster

WEIGHT
Cows 1,200 lb (550 kg);
bulls 2,000 lb (930 kg)

HEIGHT
Cows 47 in. (120 cm);
bulls 59 in. (150 cm)

**FEMALE FERTILITY/
CALVING AGE**
First calving at
22–24 months

DISTRIBUTION
Asia, Africa, Latin America,
South America, Middle East,
Pacific Islands

MAIN USE
Beef

FACT
1–2-year-old steers produce
very lean carcasses; marbling
quality is among the highest
for *Bos indicus* breeds.

COUNTRY OF ORIGIN Australia

This breed was developed in Australia in North Queensland in the 1900s by a collaborative team of cattlemen. Due to the high temperatures, the farmers needed animals that could tolerate the heat. By crossing British breeds such as the Beef Shorthorn and later the Hereford with American-bred Brahman they achieved the relevant traits. The Droughtmaster has become renowned for its resistance to ticks and being able to grow well off pastures even when nutritional levels are low, and is now a 50:50 mix of *Bos indicus* and *Bos taurus*. The low-maintenance digestive system common to *Bos indicus* is present in this breed, therefore it has a highly efficient feed conversion rate and benefits from resistance to bloat. Both males and females are docile and are sexually mature at a young age.

Droughtmasters have a distinctive look, ranging from honey to dark red in color, which helps protect them from the sun. They tend to have large ears, recessed or hooded eyes, a moderate hump, and an extended dewlap. Their loose skin helps this bovid release heat. Many ranches today are actively breeding for a lack of horns to help reduce accidents, labor costs, and to address the welfare issues associated with dehorning. As with most cattle, their skin can be used for leather.

Chianina

WEIGHT
Cows 2,000 lb (900 kg);
bulls 3,000 lb (1,350 kg)

HEIGHT
Cows 63 in. (160 cm);
bulls 65 in. (165 cm),
oxen up to 79 in. (200 cm)

**FEMALE FERTILITY/
CALVING AGE**
First calving at
32–37 months

DISTRIBUTION
Europe, Asia, the Americas,
Australia

OTHER NAMES
Calvana, Perugina

MAIN USE
Beef, draft

FACT
Originally developed as a draft
animal but later became
renowned for beef.

One of the tallest, heaviest, and oldest beef breeds in the world, the Chianina (*kee-a-nee-na*) is thought to have first existed over 2,200 years ago. The breed was popular in Italy for being a steady and sturdy draft animal until the 1970s, and is still used in parades today. Male and female oxen are well suited for agricultural work, and are easily transportable due to their gentle temperament. There are four recognized variants of this breed: Chianina of the Valdichiana; Chianina of the Valdarno; Perugina; and the Calvana. The latter has been recognized as its own breed since 1985.

One Chianina bull named Donetto received a World Record for his weight, which reached 3,920 lb (1,780 kg). Ninety percent of the Italian animals reside in the Tuscany, Umbria, and Lazio regions, where the animals were first raised. The beef cuts produced from this breed are allowed to be certified as "5R," which signifies that it is one of the five indigenous beef breeds of Italy, alongside the Marchigiana, the Maremmana, the Romagnola, and the Podolica. The Chianina has become a firm favorite in shows throughout the world. Despite calves being born at around only 110 lb (50 kg), they grow into good sized animals, with uncomplicated natural births as the skull size is relatively small.

COUNTRY OF ORIGIN Italy

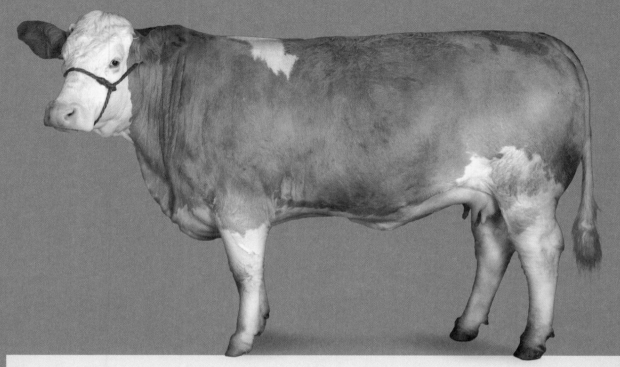

Simmental

WEIGHT
Cows 2,000 lb (900 kg);
bulls 2,900 lb (1,300 kg)

HEIGHT
Cows 55 in. (140 cm);
bulls 61 in. (155 cm)

**FEMALE FERTILITY/
CALVING AGE**
First calving at
22–24 months

DISTRIBUTION
Worldwide

OTHER NAMES
Swiss Fleckvieh, Simmentaler,
Pie Rouge

MAIN USE
Beef, dairy, traditionally draft

FACT
Can gain 3.17 lb (1.44 kg) in
weight per day.

COUNTRY OF ORIGIN Switzerland

This breed takes its name from the Swiss Valley called the Simmental, where it was first raised in the Middle Ages. With over 40 million Simmentals populating the planet, throughout all of the continents and crossed with a wide variety of other breeds, they are especially popular in Russia. Bred with several Russian breeds, the Steppe, Ukrainian, Volga, Ural, Siberian, and Far Eastern Simmental were developed. Living in Africa to Siberia, and surviving warm Alpine summers and freezing winters, this breed adapts well and is hardy. The red to brown coloring with white markings represents the traditional breed, but present day strains can have black or yellow/gold and white coats, and can be horned or polled. While able to produce high quality marbled meat, the cows can also produce 1,600 gallons (6,000 liters) of milk in their first lactation and 2,400 gallons (9,000 liters) thereafter, containing 3.7 percent protein and 4.2 percent fat.

Writers and painters have often marveled at this ancient breed, with its traditional bell around its neck on the Swiss mountains. The famous Swiss Milka chocolate brand features a purple cow based on the Simmental on its branding, and this breed was used in the original advertisements, painted purple, of course.

Tharparkar

WEIGHT
Cows 770 lb (350 kg);
bulls 900 lb (450 kg)

HEIGHT
Cows 54 in. (138 cm);
bulls 56 in. (143 cm)

**FEMALE FERTILITY/
CALVING AGE**
First calving at
38–42 months

DISTRIBUTION
Pakistan; India

OTHER NAMES
थारपारकर in Hindi, White or
Gray Sindhi (historical name),
Cutchi, Thari

MAIN USE
Milk, draft

FACT
Their milk has a fat content
of 5%.

COUNTRY OF ORIGIN Pakistan

These cattle originated in the Tharparkar District, which is now part of Pakistan, and are *Bos indicus* cattle. The Tharparkar people call them "Thari," after the nearby desert Thar, or even "Cutchi" if they live in the bordering lands of Cutch. In these regions the cattle graze upon large areas of sand dunes, which have sparse vegetation that is only available from July to September, and require feed rations from their owners the rest of the year. They have a white to gray coat but can be red and black or a combination of those colors, a black tail twitch, and a light gray to white stripe running down the spine. They have a slightly convex forehead, their long ears droop a little, and they have horns that curve upward and outward, the so called "lyre horns." Similar to other hot-climate cattle they have a medium dewlap and a large hump. They also have a strong build, making the bulls particularly suitable for draft work.

During the World War I they were relied upon to support army troops by producing milk despite the arid conditions. This breed often has Kankrej, Red Sindhi, Gyr/Gir, and Nagori breed influences. Breeders called Maldars usually keep herds of 50–300 animals, and due to the breed's more wild roaming tendencies it can be aggressive; docile if farmed.

Busa

WEIGHT
Cows 500 lb (230 kg);
bulls 900 lb (400 kg)

HEIGHT
Cows 43 in. (110 cm);
bulls 46 in. (118 cm)

**FEMALE FERTILITY/
CALVING AGE**
Sexually mature at
24 months

DISTRIBUTION
Eastern Europe

OTHER NAMES
Busha, Bosnian, Illyrian,
Bos brachyeros europeus

MAIN USE
Dairy, beef, draft

FACT
Produces a relatively good
amount of milk for its small
body size.

COUNTRY OF ORIGIN Southern Balkans/former Yugoslavia

With a long history ranging from the Neolithic period through to the present day, Busa belonged to a primitive, short-horned cattle group (*Bos brachyceros europaeus*), which has since been crossed with other breeds in neighboring regions. Busa are popular in Kosovo, and today produce most of the milk and beef in the area. Despite being able to live in mountainous terrains, purebreds of Busa are rare now, mainly limited to the mountains of Serbia. This has become a matter of concern as, although many crossbreeds are available, maintaining a pure line has become more difficult. Resistant to parasites and disease, able to graze on poor landscapes, and coping well on low management levels means that they are ideal for mountainous areas.

The Busa's coat color differs depending on the strain, which in turn relates to location. Those from the former Yugoslavia are red, the Macedonian strain is blue-gray, and three colors of black, red, and gray are found in Serbia. Considering that female calves are born weighing around 33–49 lb (15–22 kg), grow to only 397–595 lb (180–270 kg), and have lactation periods that last only 240–280 days, they produce a reasonable amount of milk. The small size makes the Busa convenient for smaller farms and smallholdings.

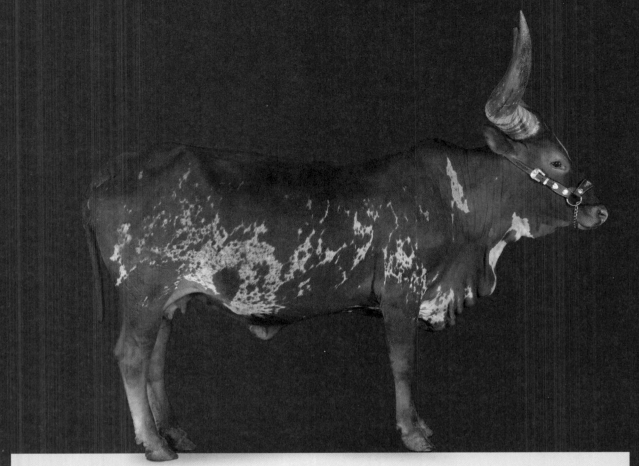

Ankole-Watusi

WEIGHT
Cows 1,100 (500 kg);
bulls 1,300 lb (600 kg)

HEIGHT
Cows 65 in. (165 cm);
bulls 78 in. (198 cm)

**FEMALE FERTILITY/
CALVING AGE**
First calving at
27–33 months

DISTRIBUTION
USA, Europe, Africa,
Australia, South America

OTHER NAMES
Ankole Longhorn, Cattle
of Kings, Royal Ox

MAIN USE
Multipurpose

FACT
Low-fat, low-cholesterol beef.

Ankole cattle originated from Africa, and some were imported from Europe to the USA in the 1920s and 30s. In 1960 the Canadian Bull was bred with the Ankole in New York State, and the breed was officially registered in 1983 at a meeting in Denver, Colorado. Therefore, although a relatively recent breed in some areas of the world, the original Ankole are thought to have originated around 4000 BCE. Indeed, they are depicted on pyramid walls, and other Egyptian art and ancient cave paintings show the breed in Africa as a sacred animal.

This breed is well known for its impressively large horns, which have the biggest circumference within cattle breeds—the largest recorded bull horns were 40¾ in. (103.5 cm). Their coat color is usually red/red and white but can be a range of colors including brown, black, dun, white, yellow, gray, and brindled. The cattle are good for milk, meat, and draft work, and their milk contains high levels of fat, around 10 percent. Ankole-Watusi calves usually have a low birth weight, which makes them ideal for breeding to heifers. This hardy breed is able to survive on relatively low levels of food and water, and adapt well to differing climates.

COUNTRY OF ORIGIN East Africa

Dexter

WEIGHT
Cows 715 lb (325 kg);
bulls 990 lb (450 kg)

HEIGHT
Cows 39 in. (100 cm);
bulls 43 in. (108 cm)

**FEMALE FERTILITY/
CALVING AGE**
Age at first calving
24–27 months

DISTRIBUTION
Africa, Australia, Europe,
New Zealand, the Americas

OTHER NAMES
The Poorman's Cow, Kerrys

MAIN USE
Beef, dairy, draft, as
companion/petting animals

FACT
The Dexter produces milk on
a par with the Jersey cow.

COUNTRY OF ORIGIN Ireland

The Dexter may well be handsome, but it is smaller than most beef or dairy breeds. Developed in Ireland and documented in 1845, it is named after Mr. Dexter, who undertook a lot of the early breed development and promotion of these cattle. At around half the size of a Hereford, it is the smallest European breed. Generally black, red, and dun in color, most have horns, but a naturally polled strain was developed in the 1990s. Due to its compact size and gentle nature, the Dexter has become an important breed in petting farms, zoos, and smallholdings, but it is also a fine production animal. With high protein and butterfat levels, the commercial Dexter can produce 1,080 gallons (4,080 liters) of milk per lactation. The meat is especially valued for its good taste and small joints.

Once classified as rare after nearly becoming extinct, the Dexter has enjoyed a resurgence and is now a recovering breed. Dwarfism (Chondrodysplasia) can affect the breed; dwarf animals have even shorter legs and can be up to 8 in. (20 cm) shorter than nonaffected Dexters. Breeding two affected animals results in loss of the calf in a quarter of the pregnancies, so is not advised.

Brahman

WEIGHT
Cows 1,300 lb (600 kg);
bulls 2,000 lb (900 kg)

HEIGHT
Cows 51 in. (130 cm);
bulls 53 in. (135 cm)

**FEMALE FERTILITY/
CALVING AGE**
First calving at
24–36 months

DISTRIBUTION
Worldwide

OTHER NAMES
American Brahman, Brahma

MAIN USE
Beef, riding (steers),
traditional sports

FACT
The beef has a uniform
distribution of low levels of
fat and high muscle content.

Named after Hindu priests called Brahmins, this breed was originally developed in the USA from 1885 using *Bos indicus* cattle from India, including the Gujarat, Ongole, Gir/Gyr, and Krishna breeds. Brahman have a large hump over their shoulders and neck, and come in a multitude of colors ranging from light gray to red/nearly black. Animals including *Bos taurus taurus* breeds imported from various countries have been used to develop Brahman further.

This breed is suited to tropical climates, as it is heat and parasite resistant, dealing well with the sun and high humidity levels, and it can also tolerate very cold temperatures. Its loose skin and specialized sweat glands help it to keep cool and its body produces less heat than other breeds. Brahman and Brahman crosses now make up over 50 percent of Australia's national herd population. It has been used for crossbreeding for many years, helping to develop other beef breeds. Brahman are known for their good mothering skills and high libido. Males are popular for riding as they are fairly docile and intelligent, in addition to being muscular. Due to their strength, they have been used for traditional bull-butting in Oman and Fujairah in the United Arab Emirates.

COUNTRY OF ORIGIN USA

Vaynol

WEIGHT
Cows 715 lb (325 kg);
bulls 935 lb (425 kg)

HEIGHT
Small to medium sized
(under 54 in./137 cm)

**FEMALE FERTILITY/
CALVING AGE**
First calving from
4–5 years

DISTRIBUTION
UK

MAIN USE
Beef, leather, conservation
grazing

FACT
Although a beef breed, Vaynol
grow relatively slowly and
mature late.

The herd originates from Scotland, and from 1872 lived in Vaynol Park, a country estate in Wales, but were moved in 1980 after their owner Sir Michael Duff passed away. They have traditionally roamed wild and are presently semi-feral, and are still not very tame, despite increased human interaction. Similar to the White Park Cattle at Chillingham with white or black coats, sometimes the coat can be white with black spots, and they have black ears and a sloping rump; the males have upswept horns.

Classified as "critical" on the Rare Breeds Survival list, this is one of the rarest British breeds, with just 150 animals remaining in three herds. Conservation efforts have ensured careful owners of the remaining stock and include a carefully managed breeding program, which uses semen from deceased bulls. The herds are kept in Yorkshire, Lincolnshire, and Scotland, to try and mitigate disease threats. The Rare Breeds Survival Trust has hopes of creating a herd back in Wales, but with the numbers of breeding stock very low, herd numbers may need to be stabilized first. In 2018 news reporters suggested that as few as 12 females were still breeding. This small, primitive breed is hardy, and perfect for conservation grazing.

COUNTRY OF ORIGIN UK

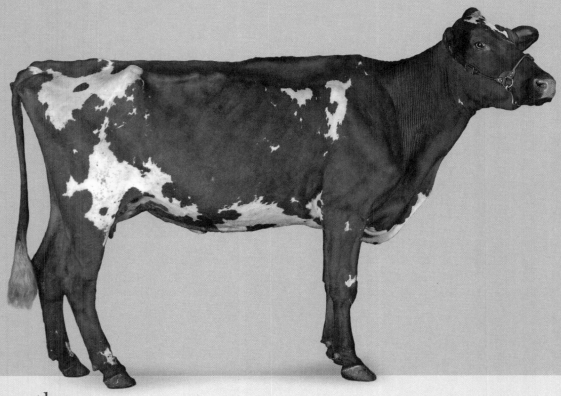

Shorthorn

WEIGHT
Dairy cows 1,300 lb
(600 kg), beef cows
1,800 lb (800 kg); bulls
2,200 lb (1,000 kg)

HEIGHT
Cows 51 in. (130 cm);
bulls 55 in. (140 cm)

**FEMALE FERTILITY/
CALVING AGE**
First calving at
24–30 months

DISTRIBUTION
Europe, the Americas, Africa,
Australia, New Zealand,

OTHER NAMES
Whitebred Shorthorn,
Durham Shorthorn

MAIN USE
Beef, dairy

FACT
Beef stock provides well
marbled meat.

COUNTRY OF ORIGIN England

The original shorthorns were bred in the North of England in the 1700s, but over the years have been selected for either dairy or beef traits, which saw the emergence of the Beef Shorthorn and the Milking Shorthorn. By 1822 the first herd book had been established, and 24 years later the American Shorthorn book was started. The more rare Northern Dairy Shorthorn are used for cheese and butter production, and are smaller than the Milking Shorthorns, at around 1,200 lb (550 kg) per cow, but still produce around 1,200 gallons (4,500 liters) of milk per lactation, containing 3–3.25 percent protein and 3.5–4.5 percent butterfat.

While all Shorthorns are red, white, or roan, the Whitebred Shorthorns have only white coats. Breeding throughout the world has produced further strains such as Poll Shorthorn and the Australian Shorthorn. The genetic condition Tibial hemimelia presents itself via a recessive inherited gene in this breed, causing severe skeletal deformities in calves; for welfare purposes euthanasia must be employed in these cases. Calves are born small for beef breeds, at around 84 lb (38 kg), and 95 percent of births need no assistance. With males having a high libido, low lameness levels in the cattle, and few cases of mastitis, this breed makes excellent breeders.

Bargur

WEIGHT
Cows 550 lb (250 kg);
bulls 660 lb (300 kg)

HEIGHT
Cows 45 in. (115 cm);
bulls 46 in. (118 cm)

**FEMALE FERTILITY/
CALVING AGE**
First calving at
45 months

DISTRIBUTION
India

OTHER NAMES
Semmarai, Bargur buffalo

MAIN USE
Milk, dung, urine, draft

FACT
Can provide around 2,900 lb
(1,300 kg) of milk per
lactation, but usually averages
around 770 lb (350 kg).

COUNTRY OF ORIGIN India

Originating in the Bargur forest in Western Tamil Nadu, in India, these cattle have fierce temperaments. Due to the rugged terrain they inhabit, they are tough animals able to endure hardship, and are small but fast. The local people believe that the Bargur's milk has healing properties, and the Lingayat people, who practice Hinduism, rear them in herds. Cattle are important within their culture and religion; therefore they are strict vegetarians, and the *ishtalinga* that they wear (usually in the form of a necklace) symbolizing a god are made from stone and cow-dung ash.

The numbers of this breed have declined rapidly, and local conservation efforts were started in 2007 following a purebred count of only 2,500 animals. In 2019 a milk cooling center was established to support milk sales. The animals are mostly housed in semi-wild conditions, in forest-based pens called "pattys." They are cared for by "Lingaiys," local tribal laborers, under zero-input conditions, being left to graze for their food. The brown/red and white cattle rarely have solely brown or white coats; both bulls and cows have horns and an ox-like hump on their neck. Tough and compact, they are certainly strong enough for draft work, and their milk, dung, and urine are much sought after.

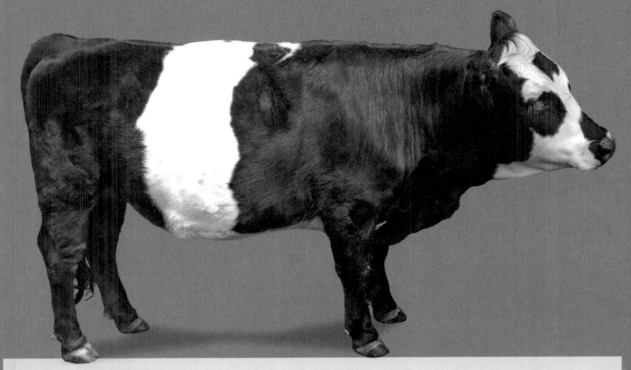

Miniature Panda

WEIGHT
Cows and bulls:
300–500 lb (135–227 kg)

HEIGHT
Under 42 in. (106 cm)

**FEMALE FERTILITY/
CALVING AGE**
First calving may occur at
24 months, but a panda
cow may not result from
the breeding

DISTRIBUTION
Mostly USA

OTHER NAMES
Panda cow

MAIN USE
Show/pet cattle, some bulls
are used for junior rodeo

FACT
The rarity of this mini cattle
means they are worth in
excess of $30,000.

COUNTRY OF ORIGIN USA

This adorable breed quite simply looks like a panda, and sadly it is even more endangered than its namesake. The first heifer born was from a miniature bull that was 75 percent Irish Dexter and 25 percent Belted Galloway bred to an exotic "Happy Mountain" cow. In 2011 there were thought to be just 24 remaining cattle alive, this has possibly risen to 40. Two more recent additions to the worldwide population include Peanut (male) and Star (female), born in Roy, Washington. The breed even appeared in *Time* magazine and the press worldwide, when Ben the bull was born on New Year's Eve 2010, in Colorado, and more have been born since. A genetic mutation gives the white belted coat coloring around the middle and the white face with black hair around the eyes.

These miniature breeds can cope well on about one acre of land per two animals, making them ideal for homesteads, smallholdings, and petting zoos and farms. Their feed conversion rates are usually around 25 percent better than their larger counterparts, making them cost efficient to keep. These are difficult animals to breed, especially in relation to achieving the standard panda look.

Randall

WEIGHT
Cows 880 lb (400 kg);
bulls 1,400 lb (650 kg)

HEIGHT
Small to medium, variable

**FEMALE FERTILITY/
CALVING AGE**
Later to mature than
most breeds

DISTRIBUTION
Canada

OTHER NAMES
None

MAIN USE
Dairy, beef, draft

FACT
Beef quality can vary across
this breed, from well marbled
through to a lean carcass with
yellow fat.

COUNTRY OF ORIGIN USA

Originally developed in Vermont, USA, on the Randall family farm in the 1900s, this breed is very rare, in part because for 80 years the family kept a closed herd, which officially became a herd in the 1990s. Once Everett Randall passed away, the cattle were sold but sadly not maintained well, until city-dweller Cynthia Creech stepped in to buy the remains of the herd and preserve the breed. Cynthia moved out to a small farm that her mother had used to keep cattle on; by this stage the cattle were down to five cows, four heifers, two yearling bulls, one weanling bull, two other calves, and one herd bull, and many were in a poor condition. Dr. Phil Sponenberg, a geneticist, came to see the cattle and designed a breed-specific breeding program. Although still on the critically endangered list, there are more than 250 breeding females today.

The Randall is variable in size and color, but usually has a white base coat with black markings. It adapts well to different environments but is particularly suited to smaller farms, smallholdings, and homesteads where it can forage for food. It is said to be a very intelligent breed, with good maternal instincts.

Appendices

Glossary

AUROCH: The common ancestor of all modern cattle.

BEEF CATTLE: Used in the meat industry.

BULL: An adult male who has not been castrated.

CALF (plural *calves*): A young female or male who has not yet been weaned off milk.

CATTLE: Generalized term for cattle (others include bovines, bulls, cows, calves). Cattle is plural, but one head of cattle can mean just one individual. Cattle generally refers to males and/or females, young or old.

COW: Once a female has given birth she becomes known as a cow; in some regions this name is given after she has had two calves. Can be used very generally to mean all cattle; it is widely accepted as there are more cows than bulls in general, especially in the dairy industry.

DEHORNED/DISBUDDED: Animals that have had horns removed.

FIRST-CALF HEIFER: A young female who has only had one calf.

FREEMARTIN: If both a female and a male twin are produced, the female will often be infertile and develop both female and male attributes.

FRESH COW: Heifer or cow who has just given birth.

HEIFER: A female who is under 3 years old and has not given birth to a calf.

INDICINE: Also called *zebu* cattle, *Bos primigenius indicus*, *Bos indicus*, or *Bos taurus indicus*, this humped cattle originates from Asia (as opposed to taurines from Europe).

MILKING/DAIRY CATTLE: Cattle used for milk production. A cow may also be called a milker or house cow, the latter might be used to refer to a single household cow.

OX (plural *oxen*): Usually a castrated male, but can be a female or bull, that is used for riding or draft work. This word also often covers domesticated and wild species such as the yak and musk ox.

POLL/POLLARD/POLLED: Cattle naturally without horns.

SPRINGER: The name given to a cow or heifer near calving.

STEER/BULLOCK: A male that has been castrated is called a steer in the USA. The word bullock changes depending on the region: in the USA it means a young bull; in other parts of the world it refers to an older steer.

TAURINE: *Bos taurus taurus*, also known as European cattle.

VEAL CALF: A calf from a dairy breed that will be used for veal meat.

WEANER: Once cattle are off milk; they become feeder calves or feeders until they are around a year old.

YEARLING/STIRK: Cattle between 1 and 2 years old.

In addition to the general English-language terms (which are widely translated into other languages), there is some terminology/slang which is specific to regions or countries:

CRITTER (USA and Canada): May be used to mean a calf or weaner.

DOGIE (USA): An orphaned calf, typically on a ranch.

MAVERICK (USA and Canada): Male or female that is unbranded.

MICKY (Australia): A wild, young bull that is unmarked. This can become a Piker bullock if caught, castrated, but lost again.

JAPANESE OX (Australia): A grain-fed steer weighing 1,100–1,400 lb (500–650 kg) who will be sold to the Japanese meat trade.

STAG/RIG (Australia, Canada, New Zealand, and other countries): A bull that is castrated late or incompletely, thereby becoming a coarse steer.

WORKING STEER (North America): Draft cattle under 4 years of age.

Bibliography

BOOKS

Budras, K.D., Habel, R.E. (2003) *Bovine Anatomy: An Illustrated Test*. Schlutersche, Germany.

Fischer, R. (1995) *Permanent Farming Systems based on Animal Traction*. Vieweg+Teubner Verlag, Germany.

Frandson, R.D., Wilke, W.L., Fails, A.D. (2009) *Anatomy and Physiology of Farm Animals*. Wiley-Blackwell, UK.

Green, M. (2012) *Dairy Herd Health*. CABI, UK.

Hotton, N. III, MacLean, P.D., Roth, J.J., Roth, E.C., eds. (1986) *The Ecology and Biology of Mammal-like Reptiles* Smithsonian Institution Press, Washington.

Robertson, J., Vickers, M. (2013) *Better Cattle Housing Design*. Agriculture and Horticulture Development Board, UK.

Sanchez, M.F. (2009) *The Estrus Cycle of the Cow: A Photographic Atlas*. Servet, Spain.

van Vuure, C. (2005) *Retracing the Aurochs—History, Morphology, and Ecology of an Extinct Wild Ox*. Pensoft Publishers, Bulgaria.

JOURNALS & REPORTS

AHDB (2018) Use of vaccines in dairy and beef cattle production 2011–2017. Agriculture and Horticulture Development Board, UK. www.farmantibiotics.org

Anderson, D.E., and St. Jean, G. (2008) Management of fractures in field settings. *Vet Clin Food Anim*. 24, pp.567–582.

Archer, S.C., Newsome, R., Dibble, H., Sturrock, C.J., Chagunda, M.G.G., Mason, C.S., Huxley, J.N. (2015) Claw length recommendations for dairy cow foot trimming. *Veterinary Record*. 177, 222.

Ballarin, C., Povinelli, M., Granato, A., et al. (2016) The brain of the domestic *Bos taurus*: weight, encephalization and cerebellar quotients, and comparison with other domestic and wild *Cetartiodactyla*. *PLoS One*. 11, 4.e0154580.

Beasley, M.J., Brown, W.A.B., and Legge, A.J. (1993) Metrical discrimination between mandibular first and second molars in domestic cattle. *International Journal of Osteoarchaeolotfy*. 3, pp.303–314.

Bhakat, M., Mohanty, T.K., Singh, S., Gupta, A.K., Chakravarty, A.K., and Singh, P. (2015) Influence of semen collector on semen characteristics of murrah buffalo and crossbred bulls. *Advances in Animal and Veterinary Sciences*. 3, 4, pp.253–258.

Bibi, F. (2007) Origin, paleoecology, and paleobiogeography of early Bovini. *Palaeogeography, Palaeoclimatology, Palaeoecology*. 248, 1, pp.60–72.

Booth, N., Briscoe, M., and Powell, R. (2000) Suicide in the farming community: methods used and contact with health services. *Occupational and Environmental Medicine*, 57, pp.642–644.

Byard, R.W. (2017) Farming deaths—an ongoing problem. *Forensic Sci Med Pathol* 13, 1, pp.1-3.

Cao, Y., Lu, Z., and Liu, Z. (2016) Foot-and-mouth disease vaccines: progress and problems. *Expert Rev Vaccines*. 15, 6, pp.783–9.

Decker, J.E., McKay, S.D., Rolf, M.M., Kim, J. et al. (2014) Worldwide patterns of ancestry, divergence, and admixture in domesticated cattle. *PLOS Genetics*. 10, 3.

Dwight, E.S.P.E., and Cannon, C. Y. (1940) The length of the intestine of calves and its bearing on the absorption of the nutrients from chyme. Journal Paper No. J-775 of the Iowa Agricultural Experiment Station, Iowa.

Eberhardt, B.G., Satrapa, R.A., Capinzaiki, C.R., Trinca, L.A., Barros, C.M. (2009) Influence of the breed of bull (*Bos taurus indicus* vs. *Bos taurus taurus*) and the breed of cow (*Bos taurus indicus*, *Bos taurus taurus* and crossbred) on the resistance of bovine embryos to heat. *Animal Reproduction Science*. 114,1–3, pp.54–61.

Edwards, C.J., Magee, D.A., Park, S.D.E., McGettigan, P.A., Lohan, A.J., et al. (2010) A complete mitochondrial genome sequence from mesolithic wild aurochs (*Bos primigenius*). *PLoS ONE*. 5,2.

Farm Animal Welfare Council/Farm Animal Welfare Committee. Five Freedoms. Archived from the original on 2012-10-07. webarchive. nationalarchives.gov.uk

Grandin, T. (2001) Cattle vocalizations are associated with handling and equipment problems at beef slaughter plants. *Applied Animal Behavior Science*. 71, pp.191–200.

Green, L.E., Huxley, J.N., Banks, C., and Green, M.J. (2014) Temporal associations between low body condition, lameness, and milk yield in a UK dairy herd. *Preventive Veterinary Medicine*. 113, 1, pp.63–71.

Groves, C.P. (1981) Systematic relationships in the Bovini (Artiodactyla, Bovidae). *Z. zool. Systematik Evol.-fschg*. 19, pp.264–278.

Gutiérrez-Gil, B., Wiener, P., and Williams, J.L. (2007) Genetic effects on coat color in cattle: dilution of eumelanin and phaeomelanin pigments in an F2-Backcross Charolais x Holstein population. *BMC Genetics*. 8, 56.

Hämäläinen, P., Takala, J., and Kiat, T.B. (2017) Global estimates of occupational accidents and world-related illnesses. Ministry of Social Affairs and Health, Workplace Safety and Health Institute, Finland. www.icohweb.org

Hassanin, A.; Douzery, E.J.P. (1999) Evolutionary affinities of the enigmatic saola (*Pseudoryx nghetinhensis*) in the context of the molecular phylogeny of Bovidae. *Proceedings of the Royal Society B: Biological Sciences*. 266, 1422, pp.893–900.

Johnston, A.M., and Edwards, D.S. (1996) Welfare implications of identification of cattle by ear tags. *Veterinary Record*. 138, 25, pp.612–4.

Kirkwood, R.M., Remnant, J.G., Payne, R.M., Murphy, A.M., and Wapenaar, W. (2018) Risk of iatrogenic damage to the sciatic nerve in dairy cattle. *Veterinary Record*. 182, 140.

Knight-Jones, T.J., and Rushton, J. (2013) The economic impacts of foot and mouth disease—what are they, how big are they, and where do they occur? *Preventive Veterinary Medicine*. 112, 3–4, pp.161–173.

Kovács, L., Kézér, F.L., Tőzsér, J., Szenci, O., Póti, P., and Pajor, F. (2015) Heart rate and heart rate variability in dairy cows with different temperament and behavioral reactivity to humans. *PloS One*. 10, 8.

Lopez-Rios, J., Duchesne, A., Speziale, D., Andrey, G., Peterson, K.A., Germann, P., Ünal, E., Liu, J., Floriot, S., Barbey, S., Gallard, Y., Müller-Gerbl, M., Courtney, A.D., Klopp, C., Rodriguez, S., Ivanek, R., Beisel, C., Wicking, C., Iber, D., Robert, B., McMahon, A.P., Duboule, D., Zeller, R. (2014) Attenuated sensing of SHH by Ptch1 underlies evolution of bovine limbs. *Nature*. 511, 7507, pp.46-51.

Lori Marino, L., and Allen, K. (2017) The psychology of cows. *Animal Behavior and Cognition*. 4, 4, pp.474–498.

Maheshwari, A., Fischer, M., Gambetti, P., Parker, A., Ram, A., Soto, C., Concha-Marambio, L., Cohen, Y., Belay, E.D., Maddox, R.A., Mead, S., Goodman, C., Kass, J.S., Schonberger, L.B., Hussein, H.M. (2015) Recent US case of Variant Creutzfeldt-Jakob Disease—global implications. *Emerg Infect Dis*. 21, 5, pp.750–9.

Marshall, F. (1989) Rethinking the role of *Bos indicus* in Sub-Sahara Africa. *Current Anthropology*. 30, 2, pp.235–240.

Meen, G.H., Schellekens, M.A., Slegers, M.H.M., Leenders, N.L.G., Erp-van der Kooij, E., and Noldus, L.P.J.J. (2015) Sound analysis in dairy cattle vocalization as a potential welfare monitor. *Computers and Electronics in Agriculture*.118, pp.111–115.

Mort, M., Convery, I., Baxter, J., and Bailey, C. (2005). Psychosocial effects of the 2001 UK foot and mouth disease epidemic in a rural population: qualitative diary based study. *British Medical Journal*, 331, 7527, pp.1234.

Nepstad, D., McGrath, D., Stickler, C., Alencar, A., Azevedo, A., and Swette, B., et al. (2014) Slowing Amazon deforestation through public policy and interventions in beef and soy supply chains. *Science*. 344, 6188.

Newsome, R.F., Green, M.J., Bell, N.J., Chagunda, M.G.G., Mason, C., Rutland, C.S., Sturrock, C.J., Whay, H.R., and Huxley, J.N. (2016) Linking bone development on the caudal aspect of the distal phalanx with lameness during life. *Journal of Dairy Science*. 99, 6, pp.4512–4525.

Okello, W., Muhanguzi, D., Macleod, E., Welburn, S., Waiswa, C., and Shaw, A.P. (2015) Contribution of draft cattle to rural livelihoods in a district of southeastern Uganda endemic for bovine parasitic diseases: an economic evaluation. *Parasites & Vectors*. 8.

Reincke, K., Saha, A., and Wyrzykowski, Ł. (2018) The global dairy world 2017/18—results of the IFCN Dairy Report 2018. International Farm Comparison Network. ifcndairy.org

Sankar, R., and Archunan, G. (2004) Flehmen response in bull: role of vaginal mucus and other body fluids of bovine with special reference to estrus. *Behav Processes*. 67, 1, pp.81–6.

Siddiq, M.K., Khan, M.A., and Akhtar, M. (2016) Bos (Mammalia: Bovidae) from the Pinjor Formation of Sardhok, Pakistan. *Pakistan Journal of Zoology*. 48, 4, pp.1071–1075.

Simpson, S., Rutland, P., and Rutland, C.S. (2017) Genomic insights into cardiomyopathies: a comparative cross-species review. *Vet Sci*. 4, 1, p19.

Sinding, Mikkel-Holger S., Gilbert, M., Thomas P. (2016) The draft genome of extinct European aurochs and its implications for de-extinction. *Open Quaternary*. 2, p7.

Takeda, K., et al. (April 2004) Mitochondrial DNA analysis of Nepalese domestic dwarf cattle Lulu. *Animal Science Journal*. 75, 2, pp.103–110.

Tatum, J.D. (2011) Animal age, physiological maturity, and associated effects on beef tenderness White Paper. Cattlemen's Beef Board and National Cattlemen's Beef Association, USA. www.beefresearch.org

The Genome Sequencing and Analysis Consortium, Elsik, C.G., Tellam, R.L., and Worley, K.C. (2009) The genome sequence of taurine cattle: a window to ruminant biology and evolution. *Science*. 324, 5926, pp.522–528.

Tullo, E., Finzi, A., and Guarino, M. (2019) Review: Environmental impact of livestock farming and Precision Livestock Farming as a mitigation strategy. *Science of The Total Environment*. 650, 2), pp.2751–2760.

United Nations Industrial Development Organization (2010) Future trends in the world leather and leather products industry and trade. UNIDO. Vienna. leatherpanel.org

Val-Laillet, D., Guesdon, V., von Keyserlingk, M.A.G., de Passillé, A.M., and Rushen, J. (2009) Allogrooming in cattle: relationships between social preferences, feeding displacements and social dominance. *Applied Animal Behavior Science*. 116, 2–4, pp.141–149.

Watts, J.M., and Stookey, J.M. (2000) Vocal behavior in cattle: the animal's commentary on its biological processes and welfare. *Applied Animal Behavior Science*. 67, 1–2, pp.15–33.

Wolf, J., Asrar, G.R., and West, T.O. (2017) Revised methane emissions factors and spatially distributed annual carbon fluxes for global livestock. *Carbon Balance and Management*. 12, 16.

Zong, G. (1984) A record of *Bos primigenius* from the Quaternary of the Aba Tibetan Autonomous Region, *Vertebrata PalAsiatica*. 22, 3. pp.239–245.

Index ❧

Author Biography

Catrin Rutland is an Associate Professor of Anatomy and Developmental Genetics at the University of Nottingham, School of Veterinary Medicine and Science, in the UK. Catrin gained her degree in Applied Biology at Derby University, studied for her Masters at the University of Nottingham and the Technical University Munich, Germany, before completing both her PhD and Medical Education Masters at Nottingham.

Catrin's research has been recognized worldwide and concentrates mainly on cardiovascular disorders such as cardiomyopathies and cancer, but also includes a wide range of anatomical studies. Her work with cattle centers on teaching bovine anatomy and researching lameness, reproduction, and cardiovascular problems. She incorporates herd-level investigations with techniques such as microscopic analysis of cells, delving into genetics, and MRI and CT scanning to fully understand cattle and their bodies.

In addition to over two hundred scientific abstracts, book chapters, books, and scientific papers, Catrin has co-authored two other "A Natural History" books: *The Horse* and *The Chicken*. She has also written scientific papers for young people in the journal *Frontiers for Young Minds*. Catrin is a British Science Association Media Representative working with Sky News and has had a number of scientific news pieces published. She was the young representative and committee member of the European Association of Veterinary Anatomists for six years and helped produce international online museums of anatomy.

While teaching undergraduate veterinary students and supervising Masters and PhD students keep her busy, Catrin writes science fiction in her spare time and enjoys giving talks about her research to the public. She lives in Derby, England, with her partner Andrew and their cat.

Acknowledgments

Writing this book has been a journey of discovery, being able to immerse myself in writing about cattle on a daily basis. It would not have been possible without the help of many people along the way. The team at Quarto Publishing, Ivy Press, The Bright Press, and Princeton University Press have been fantastic, including Katie Crous, Joanna Bentley, Tom Kitch, James Lawrence; reviewer J. A. Spencer; the brilliant illustrator John Woodcock and graphics team Jane and Chris Lanaway; and everyone else involved in the production of this book. We have all persevered through the pandemic lockdowns; your guidance, expertise, editing, and stunning artwork have been invaluable.

Working with animals is such a joy, but especially when you have fantastic students and colleagues. I would like to thank the Herd Health Group at Nottingham Vet School for being such an inspiration in all things cattle. Also thanks to Professors Samir El-Gendy, Amira Derbalah, and Karam Roshdy, Alexandria University, Egypt, for their vast knowledge on buffalo biology.

Thank you to my friends who have supported me through this process and understood why I asked them, and their children, "What would you like to know about cows?" The hundreds of responses—ranging from "Can cows jump over the moon?" and "Do cows sleep?" to "How do they produce so much milk?"—have inspired me.

Finally, thank you to my amazing partner Andrew, my parents Rhian and Paul, my brother Philip and sister-in-law Sharon, nephew Joshua, niece Erin, and the rest of my and Andrew's wonderful family. Authoring this book has often been a labor of love, and their patience, endless support, and understanding have given me the space, time, and motivation to write.

Picture Credits ✥

The publisher would like to thank the following for permission to reproduce copyright material:

Alamy/ Dominic Robinson 6 M; The Print Collector 17 B; Mike P Shepherd 26 B; Science History Images 27 B; Chronicle 32 B; imageBROKER 41 T; The Book Worm 59 T; Granger Historical Picture Archive 59 B; FLPA 67 T; Arterra Picture Library 91 B; Roger Bamber 98 B; Victor Watts 106 B; Phil Arnold 119 T; Jim West 134 B; The Granger Collection 144 M; Science History Images 166 T; Jason Smalley Photography 168; Greg Balfour Evans 175. **Bradley Pullen**: 183 T. **Dreamstime**/ Andrevaladao 12 M; Markusmayer 118 T; Wrangel 18 B; Anankkml 20 B; Johannes Gerhardus Swanepoel 21 T; Isselee 21 B; Georgios Kollidas 28 T; Viorel Sima 86 B; Mona Makela 87 M B; Potapenkoi 88 Ble; Gozzoli 92 B; Anolis01 96 T; Sunthorn Viriyapan 117 TL; Katerynakon 117 TR; Howtogoto 124 M; Kushnirov Avraham 138 B; Shailesh Nanal 139 L; Denismart 145 M; Bagwold 145 T; Tarpan 149 TR; Michael Turner 156 R; Hoxuanhuong 164 T; Tatsiana Hendzel 165 B; Stuartan 171; Drcmarx 191. **Flickr**/ Biodiversity Heritage Library 14 B; Biodiversity Heritage Library 16 L; Biodiversity Heritage Library 16 R; Girolandas la Palma 178 T; Justin Baeder 210 T. **Gert van den Bosch**: 103 TL; 163; 173 T. **Getty Images**/ Sahlan Hayes/Fairfax Media 112 B; Bloomberg 128 T; Davit Hakobyan/AFP 140 T; DeAgostini 141 T; CM Dixon/ Print Collector 141 B; Fine Art Images/Heritage Images 146 T; Fox Photos 147 B; Matthew Eisman 158 M; Per-Anders Pettersson 167 R. **iStock**/ ilbusca 26 T; emholk 39 M; AzmanL 40 T; Zwilling330 42 B; Sakan Piriyapongsak 45 B; emholk 47 B; ThinkDeep 64 B; AzmanL 75 B; nicoolay 78 B; perreten 79 T; Vladimir Zapletin 84 M; Magdevski 89 M; bohemama 93 T; ablokhin 94 B; Zu_09 94 T; mikedabell 95 L; shaunl 102 B; Neyya 105 B; RyanJLane 108 M; efreet 109 T; PaulMaguire 110 M; Jevtic 111 T; Perboge 111 B; Robert Winkler 113 B; shironosov 115 T; ktmoffitt 121 T; Lakeview_ Images 122 T; leonori 129 T; gerenme 131 T; Toa55 132 T; muri30 133 B; PeopleImages 133 T; mphillips007 135 T; TerryJLawrence 143 T; AndreyKrav 146 B; benoitb 148 T; "stellalevi" 148 B; Andrei Kravtsov 150 B; simonkr 151 B; Lynn_Bystrom 152 T; holdeneye 153 T; simonmayer 156 L; martinedoucet 157 R; PicturePartners 169 T; Gannet77 170; THEGIFT777 190 T. **Jeremy Hopley & Andrew Perris**/ 3 MR, 120 B, 172, 180, 181, 185, 186, 187, 188, 192, 194, 195, 196, 197, 198, 201, 204, 205, 206, 208. **Keith Weller/USDA** 30 BR. **LACMA**: 33 T. **Library of Congress**: Detroit Publishing Company photograph collection 163 T. **©macmonagle.com**: 41 B. **Mary Evans**/ Thaliastock 140 B; Arthur Rackham 145 B. **Nature Picture Library**: Wild Wonders of Europe/Widstrand 203 T. **Seifert Belmont Reds**: 184 T. **Science Photo Library**: Stephen Ausmus/US Department of Agriculture 166. **Shutterstock**/ andrey oleynik 1 M; Eric Isselee 2 M, 3 TL, 3 BL, 4 T; Vectorgoods studio 4 B; Svietlieisha Olena 5 B; La puma 5 T; Morphart Creation 8 B; bioraven 10 B; Pavel L Photo and Video 11 B; oorka 14 MR; Agnieszka Karpinska 14 MR; Jennifer Gottschalk 14 MR; WeHaveEverything 19 B; Svietlieisha Olena 22 B; GTS Productions 23 B; Matyas Rehak 24 B; McGraw 25 B; Dunhill 28 B; Eric Isselee 29 BL; Everett Collection 29 BR; JHVEPhoto 31 BL; smileimage9 31 BR; Bill Perry 33 MR; Alrandir 34 M; defotoberg 36 B; stockphoto mania 37 T; La puma 40 T; Flaxphotos 44 B; Azamat Fisun 44 T; Photobank.kiev.ua 48 B; Eric Isselee 50 M; Vectorgoods studio 52 T, B; Worawit Waesapaeing 55 B; vlad yan 56 M; Vera KL82 57 T; Nicram Sabod 58 T; aleksandard 60 T; Clara Bastian 61 B; sanddebeautheil 61 T; La puma 64 T; Mr. SUTTIPON YAKHAM 65 M, B; bodom 66 B; Eric Krouse 71 T; Rhian Mai Hubbart 72 T; Polarpx 73 T; J. T. Chapman 74 TL; JMx Images 74 TR; Anton Havelaar 76 outer L; suchai.guai 76 L; Svietlieisha Olena 76 outer R; Moolkum 76 M; Aleks Maks 76 R; Moolkum 77 BR; Martina Simonazzi 77 BL; inventbbart 80 M; BOULENGER Xavier 82 B; Morphart Creation 82 T; Vladimir Mulder 83 B; M. Unal Ozmen 87 MT; MasterQ 87 T; Nadezhda Nesterova 87 BL; oksana2010 87 BR; Pairoj Sroyngern 88 BR; Iakov Filimonov 88 T; Nancy Beijersbergen 90 B; Landing 95 R; Eric Isselee 97 B; Andrew b Stowe 99 T; Decent 100 T; Vladyslav Starozhylov 101 T; Grey Mountain Photo 103 TR; imagestockdesign 104 TL; David Pineda Svenske 104 TR; Eric Isselee 107 B; Photoagriculture 114 B; Fazwick 116 T; Billion Photos 117 BL; nathawit immak 118 B; Anastasiia Firsova 120 T; Juice Flair 123 B; Dean Drobot 126 BL; Cergios 126 BR; Bannafarsai_Stock 127 T; R-Mac Photography 130 B; Angorius 130 T; ArtMari 136 T; Jeff Cagle 137 T; Don Mammoser 139 R; Michael Conrad 149 TL; weber santana 151 T; zhekakopylov 152 T; Frontpage 154 T; Olesya Kuznetsova 155 B; Hypervision Creative 157 L; Diane Kuhl 160 M; Dominic Gentilcore PhD 162; Zarja 165 T; mifotodigital.club 167 L; nobeastsofierce 169 B; Sunet Suesakunkhrit 176 T; ingimar einarsson 182 T; Gozzoli 189; R. Maximiliane 200 T; Marzolino 212 M; Canicula 214 T, 216 T, 218 B; Angorius 220 T; Azamat Fisun 221 B; La puma 222 B. **Unsplash**/ sippakorn yamkasikorn 9 M; Mihail Macri 10 T; Gunjan Bhattacharjee 136 B. **Wellcome Collection**: 15 T, 15 B, 142 B. **Wikimedia Commons**/ Lucinda Morgan 174 T; Homoatrox 177; Pavanaja U. B. 179 T, 202 T, 209 T; Karl Young 193 T; Cgoodwin 199 T; Claire.cojean.05 207 T; CTPhil 211 T. **Wisconsin Historical Society**: WHS-2115-1 128 B.

All reasonable efforts have been made to trace copyright holders and to obtain their permission for the use of copyright material. The publisher apologizes for any errors or omissions and will gratefully incorporate any corrections in future reprints if notified.